Lecture Notes in Electrical Engineering

Volume 1193

The book series *Lecture Notes in Electrical Engineering* (LNEE) publishes the latest developments in Electrical Engineering—quickly, informally and in high quality. While original research reported in proceedings and monographs has traditionally formed the core of LNEE, we also encourage authors to submit books devoted to supporting student education and professional training in the various fields and applications areas of electrical engineering. The series cover classical and emerging topics concerning:

- Communication Engineering, Information Theory and Networks
- Electronics Engineering and Microelectronics
- Signal, Image and Speech Processing
- Wireless and Mobile Communication
- Circuits and Systems
- Energy Systems, Power Electronics and Electrical Machines
- Electro-optical Engineering
- Instrumentation Engineering
- Avionics Engineering
- Control Systems
- Internet-of-Things and Cybersecurity
- Biomedical Devices, MEMS and NEMS

For general information about this book series, comments or suggestions, please contact leontina.dicecco@springer.com.

To submit a proposal or request further information, please contact the Publishing Editor in your country:

China

Jasmine Dou, Editor (jasmine.dou@springer.com)

India, Japan, Rest of Asia

Swati Meherishi, Editorial Director (Swati.Meherishi@springer.com)

Southeast Asia, Australia, New Zealand

Ramesh Nath Premnath, Editor (ramesh.premnath@springernature.com)

USA, Canada

Michael Luby, Senior Editor (michael.luby@springer.com)

All other Countries

Leontina Di Cecco, Senior Editor (leontina.dicecco@springer.com)

**** This series is indexed by EI Compendex and Scopus databases. ****

Maode Ma
Editor

Proceedings of the 12th International Conference on Communications, Circuits, and Systems

ICCCAS 2023, 5–7 May, Singapore

Springer

Editor
Maode Ma
KINDI Center for Computing Research
Future Institute of Engineering
and Technology
Qatar University
Doha, Qatar

ISSN 1876-1100 ISSN 1876-1119 (electronic)
Lecture Notes in Electrical Engineering
ISBN 978-981-97-2635-6 ISBN 978-981-97-2636-3 (eBook)
https://doi.org/10.1007/978-981-97-2636-3

This Springer imprint is published by the registered company Springer Nature Singapore Pte Ltd.
The registered company address is: 152 Beach Road, #21-01/04 Gateway East, Singapore 189721, Singapore

Organization

Conference Committee Chairs

Franco Maloberti, FIEE, University of Pavia, Italy
Massimo Alioto, FIEEE, National University of Singapore, Singapore

Conference Committee Co-Chair

Maode Ma, FIET, Qatar University, Qatar

Program Committee Chairs

Lin Wang, SMIEEE, Xiamen University, China
Haruo Kobayashi, SMIEEE, Gunma University, Japan

Program Committee Co-Chairs

Yang Yue, SMIEEE, Xi'an Jiaotong University, China
Jeff Kilby, Auckland University of Technology, New Zealand
Kanglian Zhao, Nanjing University, China

Award Chairs

Char-Dir Chung, FIEEE, National Taiwan University, Taiwan
Jihong Yu, Beijing Institute of Technology, China

Local Chair

Eunice Wang, Sensors and Systems Society, Singapore

Treasurer

Yuanshu Chen, University of Electronic Science and Technology of China, China

Technical Program Committees

Anh Tuan Do, Institute of Microelectronics, Singapore
Alonso Sanchez, Universidad Peruana de Ciencias Aplicadas, Peru
Ajay Gurjar, Sipna College of Engineering and Technology, India
Ayinuer·Nuertai, Xinjiang Police College, China
Ankan Bhattacharya, Hooghly Engineering and Technology College, India
Alkiviadis Hatzopoulos, Aristotle University of Thessaloniki, Greece
Basavaraj M. Angadi, Basaveshwar Engineering College, India
Chan-Tong Lam, Macao Polytechnic University, China
Christian Schindelhauer, University of Freiburg, Germany
Chuang Bi, University of Electronic Science and Technology of China, China
Cong Wang, Sichuan Police College, China
Eun-Jun Yoonm, Kyungil University, Korea
Graziano Chesi, The University of Hong Kong, China
Hailin Xiao, Guilin University of Electronic Technology, China
Hamid Ali Abed Al-asadi, Basra University, Iraq
Hiroyuki Yamauchi, Fukuoka Institute of Technology, Japan
Huang Yu-Che, Chaoyang University of Technology, Taiwan
Idawaty Ahmad, University Putra Malaysia, Malaysia
Ismail Rakip Karas, Karabuk University, Turkey
Jung-Shian Li, Taiwan National Cheng Kung University, Taiwan
Ka C. Chan, La Trobe University, Australia
Khairul Azhar Mat Daud, University Malaysia Kelantan, Malaysia
Kwen Siong Chong, Zero-Error Systems Pte Ltd, Singapore

Kelvin Xuanyao Fong, National University of Singapore, Singapore
Kuangang Fan, Jiangxi University of Science and Technology, China
Luca Marchetti, University of South-Eastern Norway, Norway
Manish Bhardwaj, KIET Group of Institutions, India
Mehul Mahrishi, SKIT, Rajasthan Technical University, India
Ming Li, North China University of Technology, China
Min Lin, Shanghai University, China
Mingqiang Huang, Shenzhen Institutes of Advanced Technology, Chinese Academy of Sciences, China
Mingyu Wang, Sun Yat-sen University, China
Mohammed M. Bait-Suwailam, Sultan Qaboos University, Oman
Mohana Sundaram Muthuvalu, Universiti Teknologi PETRONAS, Malaysia
Md. Kafiul Islam, Independent University, Bangladesh
Nipun Sharma, Presidency University, India
Norasmadi Abdul Rahim, Universiti Malaysia Perlis, Malaysia
Pakorn Keawtrakulpong, King Mongkut's University of Technology Thonburi (KMUTT), Thailand
Paulo Roberto De Lira Gondim, Universidade de Brasilia, Brazil
Prapaisri Sudasna na Ayudthya, Kasetsart University, Bangkaen Campus, Thailand
Pulak Mondal, University of Calcutta Kolkata, India
Qian Kong, University of Electronic Science and Technology of China, Zhongshan Institute, China
Shanton Chang, The University of Melbourne, Australia
Slami Saadi, Ziane Achour University of Djelfa, Algeria
Sohiful Anuar Zainol Murad, Universiti Malaysia Perlis, Malaysia
Sonam Rewari, Delhi Technological University Delhi, India
Sun Wook Kim, Electronics and Telecommunications Research Institute, Korea
Swati Sharma, Presidency University, India
Tao Wu, Shantou University, China
U. D. Dalal, Sardar Vallabhbhai National Institute of Technology (SVNIT), India
Xiaojuan Sun, Chinese Academy of Sciences, China
Xi Sung Loo, Nanyang Technological University, Singapore
Yuanjin Zheng, Nanyang Technological University, Singapore
Yu Xiaohong, Shandong Management University, China
Zulkifli Husin, Universiti Malaysia Perlis, Malaysia

Preface

The 12th International Conference on Communications, Circuits, and Systems (ICCCAS 2023) was successfully held during May 5–7, 2023 in a combination of onsite (Singapore) and online. This conference brought together experts and scholars from various fields, fostering a vibrant academic atmosphere throughout the event. We extend our heartfelt appreciation to all the attendees for their active participation and contribution to the success of this conference.

The diverse activities (keynote & invited speeches, best student competition, and parallel sessions) provided a platform for the exchange of knowledge, ideas, and research findings among the participants. The valuable insights shared during the conference undoubtedly enriched the academic discourse and inspired further exploration in our respective areas of study. We would like to express our sincere gratitude to the distinguished speakers who graced this occasion with their expertise and delivered excellent speeches.

We would also like to extend our deepest appreciation to the diligent and dedicated technical committees and reviewers who meticulously reviewed papers. The highest standards of conference proceedings cannot be separated from their expertise and fair & rigorous evaluation, which has significantly enriched the content of this conference proceeding. Thanks again for their contributions to this proceedings volume. And it's pleasant that this conference proceeding will be helpful for all readers in the area of communications, circuits, and systems and other related areas.

Thank you to all participants and all committee members for their generous support and contribution. We sincerely hope that ICCCAS will become a platform that is preferred by everyone. We will spare no effort to improve and develop ICCCAS platform. Look forward to get together again with all of you next year!

Doha, Qatar

Maode Ma
Conference Chair

Contents

Next Generation Communication Network Technology and Future Development

Electronic and Communication Engineering

Wireless Sensor Network and Wireless Communication Technology

Double Walsh-Hadamard Transform for Improved Efficiency in OFDM-IM

Yuhao Lian, Yuhua Wang, and Xinyu Lin

Abstract Amidst the rapidly evolving landscape of B5G/6G wireless networks, the emergence of index modulation has marked a significant development, and one key technique gaining ground is orthogonal frequency division multiplexing-index modulation (OFDM-IM), which boosts both communication and signal transmission spectral efficiency. Although OFDM-IM is an innovative multicarrier modulation technique that builds on the conventional OFDM system. It still confronts a significant challenge related to the High Power Amplifier (HPA) cost, which arises from the distortion that OFDM-IM signals cause when passing through nonlinear devices. This distortion is driven by the considerable Peak-to-Average Power Ratio (PAPR) that characterizes the OFDM-IM signal. In this paper, we propose a novel waveform named Double WHT-OFDM-IM, which combines Double Walsh-Hadamard Transform (WHT) with orthogonal frequency division multiplexing-index modulation (OFDM-IM) to improve the system performance and combat the enormous PAPR of WHT-OFDM-IM in the low number of activated subcarriers scenarios. The proposed approach satisfies the demanding criteria of 6th-generation (6G) and beyond communication schemes, specifically with regard to low peak-to-average power ratio (PAPR), and low bit-error rate (BER).

Keywords B5G/6G wireless networks · Low PAPR · OFDM-IM · Double WHT-OFDM-IM

Y. Lian (✉) · Y. Wang · X. Lin
Chongqing University of Posts and Telecommunications, Chongqing 400065, China
e-mail: lianyuhao@ieee.org

Y. Wang
e-mail: curley0619@gmail.com

X. Lin
e-mail: 1520454049@qq.com

M. Ma (ed.), *Proceedings of the 12th International Conference on Communications, Circuits, and Systems*, Lecture Notes in Electrical Engineering 1193,
https://doi.org/10.1007/978-981-97-2636-3_1

1 Introduction

In wireless communication systems, the relentless pursuit of higher spectrum and energy efficiency at a low cost has become a critical objective. To achieve this challenging goal, multi-carrier technology has emerged as a promising approach. By employing parallel transmission, this technology enhances spectrum efficiency within constrained spectrum resources. More specifically, it converts serial high-speed information streams into parallel low-speed information streams, which are subsequently transmitted in superposition. Orthogonal frequency division multiplexing (OFDM) is a leading example of multi-carrier technology and serves as a key waveform for both 4G LTE and 5G new radio (NR) systems. OFDM is particularly well-suited to enhanced mobile broadband (eMBB) data transmission over multicarrier bases, as it offers higher spectrum efficiency compared to other alternatives [1].

Higher spectrum efficiency and improved performance are key requirements for the upcoming B5G/6G wireless networks. To meet these demands, index modulation (IM) has emerged as a promising solution, attracting significant attention in recent years [2]. In 2009, Abu.alhiga and Hass introduced the concept of spatial modulation to the OFDM system and proposed a novel technology called Index Modulation based on OFDM (SIM-OFDM) [3]. Building on this, in 2013, Basar proposed an improved version, called Orthogonal Frequency Division Multiplexing with Index Modulation (OFDM-IM) [4], which provides greater flexibility in adjusting the number of active subcarriers and the systematic mapping, thus resolving the error propagation issue in SIM-OFDM. Research has demonstrated that OFDM-IM technology outperforms SIM-OFDM with lower bit error rates (BER), faster transmission rates, and higher spectrum efficiency. Additionally, research into low complex and efficient detector in OFDM-IM has also seen ongoing progress [5]. While OFDM-IM shows promise in addressing some of the challenges of traditional OFDM, such as inter-symbol interference and frequency-selective fading, it still faces a significant hurdle in the form of high peak-to-average power ratio (PAPR). Therefore, researchers have been exploring various strategies to mitigate this issue. Some of the techniques that have been proposed include adaptive companding [6], a novel weighted tone reservation method [7], and differential evolution [8]. These approaches aim to reduce PAPR and improve system performance by adjusting the signal's amplitude and phase, optimizing tone reservation, and leveraging evolutionary algorithms. Efforts to address PAPR and improve OFDM continue to evolve, and additional strategies are also being explored [9–11]. The goal is to enhance the overall efficiency, reliability of OFDM systems for various applications.

On the contrary, the Walsh-Hadamard Transform (WHT) has been widely utilized in digital image processing [12] and is progressively being applied for various purposes in communication systems, such as signal processing and signal precoding [13]. In 2000, WHT-OFDM was introduced as a method to significantly enhance bit error rate (BER) performance [14]. WHT-OFDM has the potential to substantially reduce the high peak-to-average power ratio (PAPR), which can lead to expensive high power amplifiers (HPA) [15]. Additionally, the use of WHT in non-orthogonal multiple access (NOMA) can greatly improve 5G communication performance [16]. Furthermore, WHT-OFDM-IM demonstrates superior performance in BER and PAPR in scenarios with multiple subcarriers and multiple activated subcarriers. Building on the foundation of WHT-OFDM-IM, this study introduces the concept of Double WHT in OFDM-IM, resulting in a novel system scheme known as Double Walsh-Hadamard Transformed OFDM-IM (Double WHT-OFDM-IM). Through comparison, Double WHT-OFDM-IM is found to significantly reduce BER and effectively mitigate the high PAPR of WHT-OFDM-IM in scenarios with a low number of activated subcarriers. This is an innovative system scheme with promising potential to enhance communication system performance.

In this paper, we tackle the challenge of high PAPR in scenarios with a low number of activated subcarriers in WHT-OFDM-IM systems and aim to enhance the performance of OFDM-IM. To achieve this, we propose a novel system scheme, Double Walsh-Hadamard Transformed OFDM-IM (Double WHT-OFDM-IM), which significantly improves communication efficiency. The main contributions of this paper are summarized as follows:

- Building upon the foundation of the WHT-OFDM-IM system, we first have made improvements to its system structure. We have proposed a novel system model, called the Double Walsh-Hadamard Transformed OFDM-IM (Double WHT-OFDM-IM), which employs Double WHT. In addition, we have provided a detailed explanation of the relevant principles based on OFDM-IM.
- We the proceeded to assess the PAPR of Double WHT-OFDM-IM in various activated subcarrier scenarios by conducting a CCDP curve comparison with OFDM-IM, WHT-OFDM-IM.
- Furthermore, we were able to visually illustrate the enhancement effect of Double WHT on communication quality through constellation observation. Meanwhile, we assessed the performance of Double WHT-OFDM-IM by comparing the BER among OFDM-IM, WHT-OFDM-IM, and Double WHT-OFDM-IM.

This paper is organized as follows: The detailed principle of the Double WHT-OFDM-IM system model is described, and the principle of Walsh-Hadamard Transform (WHT) are emphasized in Sect. 2. In Sect. 3, we obtain the transceiver structure of WHT-OFDM-IM by comparing the transceiver structure of OFDM-IM with WHT-OFDM-IM in Sect. 4. In Sect. 4, we conduct simulations to compare and analyze the PAPR and BER of Double WHT-OFDM-IM and OFDM-IM. Finally, the paper concludes with a summary in Sect. 5.

2 Double WHT-OFDM-IM Operating Principle

Double WHT-OFDM-IM is a novel and promising technique that improves upon conventional OFDM-IM and newly WHT-OFDM-IM by achieving a lower peak-to-average power ratio (PAPR) and superior performance. By selectively activating a subset of subcarriers for transmitting modulation symbols, Double WHT-OFDM-IM divides the transmitted information bits into index and symbol parts. The index bits are conveyed through the combination of activated subcarrier indices, while the symbol bits are carried by the modulation symbols transmitted on the active subcarriers. This approach offers the potential to minimize PAPR and maximize spectral efficiency, making Double WHT-OFDM-IM an attractive solution for achieving high-performance wireless communications.

Given the independence of subblocks in Double WHT-OFDM-IM, the transmitter chooses k activated subcarriers from the n subcarriers in the index mapper. Therefore, the combination of activated subcarriers for each Double WHT-OFDM-IM subblock β can be expressed as:

$$I_\beta = \left\{ i_\beta^1, i_\beta^2, \sim \ldots, \sim i_\beta^k \right\} \tag{1}$$

the Double WHT-OFDM-IM transmitter evenly divides a total of N subcarriers into g groups, with each group containing $n = N/g$ subcarriers. Additionally, an m-bit binary sequence is also evenly divided into g groups, with each group consisting of $p = m/g$ bits. The p-bit sequence is further divided into an index bit sequence p_1 and an information bit sequence p_2. The number of bits of modulation symbols carried by each Double WHT-OFDM-IM transmission subblock is:

$$\tfrac{m_2}{g} = p_2 = k\left(\log_2 M\right) \tag{2}$$

Similarly, the number of index bits carrying activated subcarrier information in each Double WHT-OFDM-IM transmission subblock can be expressed as:

$$\frac{m_1}{g} = p_1 = \left\lfloor \log_2 C(n, k) \right\rfloor \tag{3}$$

The p bits are subsequently mapped onto the Double WHT-OFDM-IM subblocks, and the g Double WHT-OFDM-IM subblocks are combined to form an Double WHT-OFDM-IM transmission block of length N [4], as illustrated in Fig. 1.

$$m = pg = k\left(\log_2 M\right) + \left\lfloor \left\lfloor \log_2 C(n,\ k) \right\rfloor \right\rfloor g \tag{4}$$

In accordance with the theory outlined above, Fig. 1b shows that when the number of subcarriers is $n = 6$ and the number of activated subcarriers is $k = 2$, the index bit p_1 of each block in the Double WHT-OFDM-IM under 128QAM is equal to 3, and the information bit p_2 is equal to 7.

To select k out of n subcarriers as activated subcarriers for each Double WHT-OFDM-IM subblock β, the transmitter utilizes the index bit sequence p_1 to map the activated subcarrier indices through the index mapper. Moreover, the M-ary modulator maps the symbol bit sequence p_2 to generate a vector of k modulation symbols. This leads to the following expression for the combination of activated subcarrier indices:

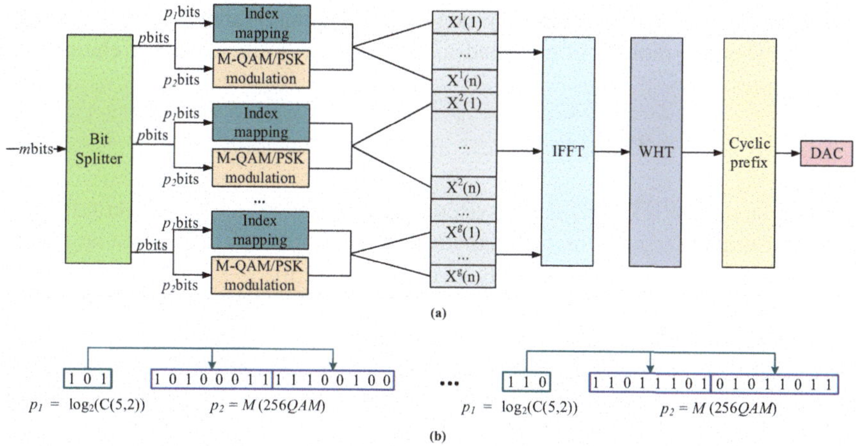

Fig. 1 Schematic illustration depicting the principle of Double WHT-OFDM-IM: **a** represents the transmitter of Double WHT-OFDM-IM, and **b** provides an example of the division between the index bit and information bit in OFDM-IM (with 6 subcarriers, 2 of which are active under 128QAM modulation)

$$S_\beta = \left\{ s_\beta^1,\ s_\beta^2,\ \ldots,\ s_\beta^k \right\} \tag{5}$$

The index mapping method in Double WHT-OFDM-IM is similar to that of OFDM-IM, which primarily employs two approaches: (1) the look-up table method and (2) the combinatorial method [14].

(1) Look-up table method: for instance, when considering $n = 5$ subcarriers and $k = 4$ activated subcarriers, all possible permutations of the active subcarriers result in $C(5{,}4) = 1$ combinations. However, in Double WHT-OFDM-IM, one permutation is abandoned, which can be observed in Table 1.
(2) Combinatorial method: This method offers a solution for mapping natural numbers and index positions in permutation $C(n, k)$ addressing the issue of a large look-up table when the index bit is extensive. As a result, index mappings are commonly computed using combinatorial methods.

Assuming the number of activated subcarriers is k, the index positions of the k activated subcarriers arranged in descending order can be represented as $J = \{c_k, \ldots, c_1\}$. For any positive integer $Z \in [0, C(n, k) - 1]$, there is a unique set of J,

Table 1 Index mapping table

$p1$ bits	Subcarrier index combination	Subblock
00	3, 2, 1, 0	$[0s^1s^2s^3s^4]$
01	4, 2, 1, 0	$[s^1 0 s^2s^3s^4]$
10	4, 3, 1, 0	$[s^1s^2 0 s^3s^4]$
11	4, 3, 1, 0	$[s^1s^2s^3 0 s^4]$

resulting in J and Z a one-to-one mapping in the combinatorial number method. For instance, when the number of activated subcarriers $k = 4$, the mapping relationship can be expressed as:

$$Z = C(c_4, \ 4) + C(c_3, \ 3) + C(c_2, \ 2) + C(c_1, \ 1) \tag{6}$$

In the Double WHT-OFDM-IM system, the index bit p^1 is first converted into a positive decimal integer Z in each subblock at the transmitter. Next, the largest c_k is found such that $C(c_k, k) \leq Z$, followed by the selection of the largest c_{k-1} value such that $C(c_{k-1}, \ k-1) \leq z - c(c_k, k)$, and so on, until all the index positions are determined. The redundant activated subcarrier permutations are then abandoned. For instance, if $n = 6$ and $k = 3$, with $C(6{,}3) = 20$, the process is illustrated as follows:

$$\begin{aligned} &19 = C(5, \ 3) + C(4, \ 2) + C(3, \ 1) \rightarrow J = \{5, \ 4, \ 3\} \\ &\ldots \\ &1 = C(3\ 3) + C(1, \ 2) + C(0, \ 1) \rightarrow J = \{3, \ 1, \ 0\} \\ &0 = C(2, \ 3) + C(1, \ 2) + C(0, \ 1) \end{aligned}$$

Referring to the aforementioned theoretical analysis, the activated subcarriers can be obtained through the index mapping of the p_1 bit. For the symbols located on the activated subcarriers, traditional M-QAM/M-PSK modulation is executed according to the information of the p_2 bit. Hence, the Double WHT-OFDM-IM symbols can be expressed as:

$$X_\beta = \left\{x_\beta^1, \ x_\beta^2, \ldots, \ x_\beta^n\right\} \tag{7}$$

Furthermore, in Double WHT-OFDM-IM, we use the IFFT algorithm to complete orthogonal operation to high-light the essence as a novel OFDM system.

$$Y_\beta = IFFTX_\beta \tag{8}$$

In contrast to conventional OFDM-IM wireless systems, both the input and output data of Double WHT-OFDM-IM need to be subjected to Walsh-Hadamard Transform (WHT) operations, both before and after the inverse fast Fourier transform (IFFT) processing. This means that at the receiver, an additional inverse WHT (IWHT) algorithm is necessary to recover the original data.

The Walsh-Hadamard Transform (WHT) is a non-sinusoidal function transform that utilizes orthogonal rectangular functions as primary functions and shares similar properties with the Fourier function. In the transmitter, WHT is achieved by multiplying the input signal with the WHT matrix, and the recovered signal is obtained in the receiver by multiplying the received signal with the same WHT matrix. Consequently, the relationship between the input and output signals can be expressed as:

$$Z = H_i Y_\beta \Leftrightarrow Y_\beta = H_i Z_\beta \tag{9}$$

where Y_β and Z_β are input and output signals respectively, H_i is WHT matrix, a precoding matrix, whose code length $i = 2^l (l \in Z^+)$. To better illustrate the properties of the WHT matrix, let us assume that H is an orthogonal matrix composed of 1 and -1. In this case, the block matrix can be expressed as follows:

$$\begin{pmatrix} 1 & 1 \\ 1 & -1 \end{pmatrix}, \begin{pmatrix} 1 & 1 & 1 & 1 \\ 1 & -1 & 1 & -1 \\ 1 & 1 & -1 & -1 \\ 1 & -1 & -1 & 1 \end{pmatrix} \cdots \begin{pmatrix} H & H \\ H & -H \end{pmatrix} \tag{10}$$

thus, the general sequence of WHT matrix can be expressed as.

$$\begin{aligned} H_1 &= (1), H_2 = \frac{1}{\sqrt{2}} \begin{pmatrix} 1 & 1 \\ 1 & -1 \end{pmatrix}, \ldots, \\ H_{i=2^l} &= \frac{1}{\sqrt{i}} \begin{pmatrix} H_{k=2^{l-1}} & H_{k=2^{l-1}} \\ H_{k=2^{l-1}} & -H_{k=2^{l-1}} \end{pmatrix} \end{aligned} \tag{11}$$

where $\frac{1}{\sqrt{i}}$ is expressed as a normalization factor, as the order of the Hadamard matrix, k is used to derive the WHT matrix.

The primary objective of the receiver is to identify the combination of activated subcarrier indices and the transmission symbols on those subcarriers. Similar to conventional OFDM-IM systems, the maximum likelihood (ML) hard decision algorithm and the log-likelihood ratio (LLR) soft decision detection algorithm are commonly utilized [4]. To maximize the performance improvement of Double WHT-OFDM-IM over OFDM-IM, this paper employs ML detection.

The core idea of the ML detection algorithm is to search and compare all possible combinations of the index $\hat{I}^i$ and modulation symbol vectors s^i of the activated subcarriers. The ML algorithm seeks to minimize the value of the following equation:

$$\left(\hat{I}^i, \hat{s}^i\right) = \arg\min_{I^i, s^i} \sum_{a=1}^{n} \left|\mathcal{Y}_a^i - \mathcal{X}_a^i\right|^2 \tag{12}$$

ere $\hat{I}^i$ is the activated index combination of each sub-block, and $\hat{s}^i$ is the modulation symbol vector. $\mathcal{Y}^i(a)$ and $\mathcal{X}^i(a)$ is the received signal and send signal respectively.

3 The Transceiver Structure of Double WHT-OFDM-IM

This section aims to delineate the system architecture of the Double WHT-OFDM-IM and establish the corresponding simulation parameters.

The transceiver structure in Double WHT-OFDM-IMIM is a fusion of the traditional OFDM-IM and WHT-OFDM-IM transceivers [4, 8]. However, Double WHT-OFDM-IM integrates WHT between M-ary modulation and IFFT, between the IFFT and the cyclic prefix in the transmitter, and includes an IWHT between the FFT and the cyclic prefix, as well as between ML detection and FFT in the receiver, as compared to the transceiver structure of OFDM. The transmitter functions primarily include: index modulation, MQAM/MPSK coding, implementation of IFFT and WHT algorithms, addition of a cyclic prefix to the output of IFFT, completion of digital-to-analog (DAC) conversion, and final output to the AWGN channel. In the receiver, as depicted in Fig. 2, the received signal is first converted to analog-to-digital (ADC) signal. The cyclic prefix is then removed, and the output is obtained through the use of the IWHT algorithm. After the IWHT algorithm, the signal is processed with FFT and IWHT algorithm again. At this stage, channel estimation and equalization operations are performed to eliminate the effects of the channel on the ML detection algorithm. In this paper, the ideal zero-forcing (ZF) equalization is adopted, given by:

$$Y_E = \frac{P_p}{P_p \cdot P} Y \tag{13}$$

where Y is received signal by FFT, P_p is the complex conjugate transpose of P.

Finally, the ML detection algorithm and the corresponding modulation format decoding are used to complete the entire receiver process.

4 Simulation Results

Combined with the simulation of the theoretical analysis in Sects. 2 and 3, the simulation parameters are obtained as shown in Table 2. If the difference of parameters is not indicated in this section, all simulation parameters refer to Table 2.

The CCDF curve comparing OFDM-IM with the WHT-OFDM-IM, Double WHT-OFDM-IM under different nu-members of activated subcarriers are shown in Fig. 3. It reveals that when the number of activated subcarriers is small, the peak-to-average power ratio (PAPR) of Double-WHT-OFDM is lower than that of WHT-OFDM-IM, effectively mitigating the problem of PAPR increase caused by a single WHT applied to OFDM-IM. Conversely, when the number of activated subcarriers in WHT-OFDM is close to the total number of subcarriers, the use of WHT can reduce the PAPR of the OFDM-IM system. Similarly, Table 3 shows the PAPR estimations comparing WHT-OFDM-IM with Double WHT-OFDM-IM under different numbers

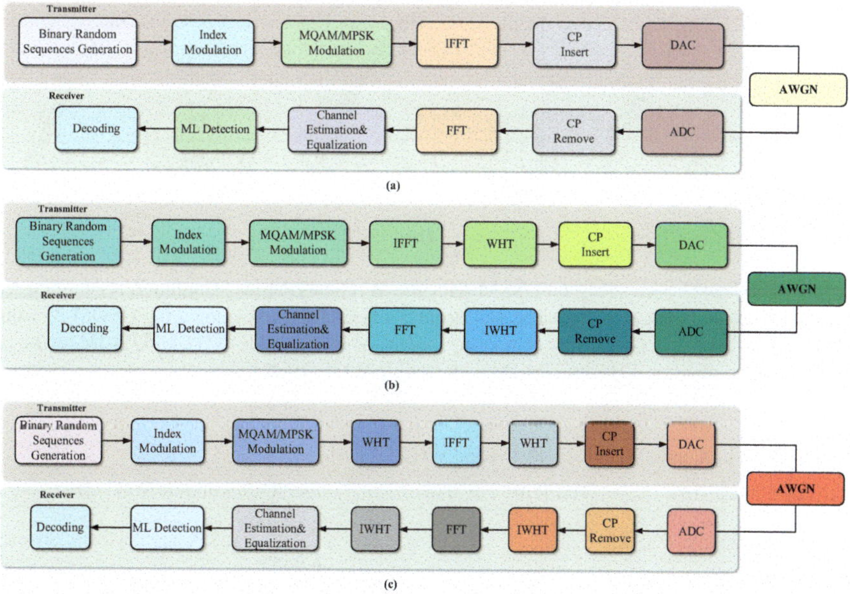

Fig. 2 Schematic of the OFDM-IM, WHT-OFDM-IM, and Double WHT-OFDM-IM transceiver architecture: **a** represents the OFDM-IM transceiver architecture, **b** illustrates the WHT-OFDM-IM transceiver architecture, and **c** depicts the Double WHT-OFDM-IM transceiver architecture

Table 2 The simulation parameters setting

Parameters	Value
Subcarrier numbers	16
Activated subcarriers	12
Length of sequence	140000
Modulation formats	64QAM
CP	0.125
quantization bit	11
Detection	ML
SNR	0 ~ 25 dB
PARP	0 ~ 12 dB

of activated subcarriers, which is consistent with the results reflected by CCDF curve. Hence, it is necessary to dynamically select the appropriate WHT scheme based on the number of active subcarriers to tackle the OFDM-IM problem.

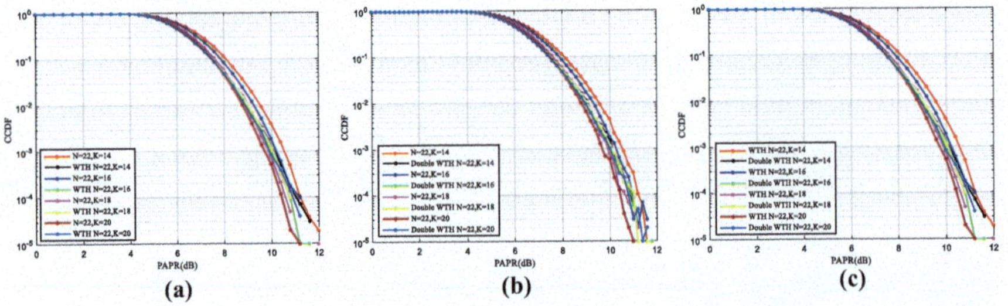

Fig. 3 CCDF curve comparing OFDM-M with the WHT-OFDM-IM, Double WHT-OFDM-M under different numembers of activated subcarriers: **a** is CCDF curve comparing OFDM-M with WHT-OFDM-IM, **b** is CCDF curve comparing OFDM-M with Double WHT-OFDM-IM, and **c** is CCDF curve comparing WHT-OFDM-IM with Double WHT-OFDM-IM

Table 3 Parp in double WHT-OFDM-IM and WHT-OFDM-IM

Activated subcarriers	PAPR after WHT/dB	PAPR after double WHT/dB
14	6.5468	**6.0630**
16	6.3740	**6.0443**
18	**6.0584**	6.2022
20	**6.0263**	6.0941

The above results indicate that WHT-OFDM-IM can considerably combat the enormous PAPR in the multiple activated subcarriers scenarios compared with OFDM-IM and Double WHT-OFDM-IM can considerably combat the enormous PAPR in the low number of activated subcarriers scenarios compared with WHT-OFDM-IIM, overcoming the common disadvantage of traditional OFDM-IM as an OFDM system. Therefore, adapting and dynamically selecting between WHT and Double WHT for different PAPR scenarios will be a novel area of future research and exploration for WHT-OFDM-IM and Double WHT-OFDM-IM.

In the AWGN channel, Gaussian white noise can have a significant impact on the received signal, as demonstrated in Fig. 4. Specifically, Fig. 4a–c show the received signal constellations of Double WHT-OFDM-IM compared with those of OFDM-IM and WHT-OFDM-IM at 10 dB SNR, while Fig. 4d–f show the received signal constellations at 15dBSNR. As demonstrated in Fig. 4, Double WHT-OFDM-IM signals are less distorted by noise than WHT-OFDM-IM and OFDM-IM, while the discernibility of received signal constellation points is significantly improved by Double WHT.

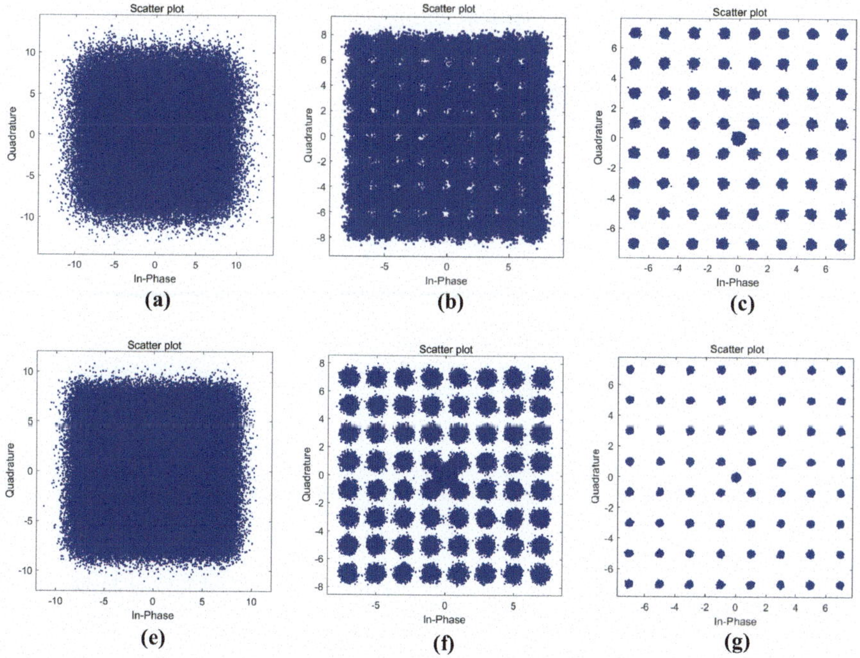

Fig. 4 Constellation of received signals in Double WHT-OFDM-IM, WHT-OFDM-IM and OFDM-IM under the same SNR: **a** is the con-stellation of received signals in OFDM-IM under 10 dB SNR, **b** is the constellation of received signals in WHT-OFDM-IM under 10 dB SNR, **c** is the constellation of recived signals in Double WHT-OFDM-IM under10dB SNR, **d** is the constellation of received signals in OFDM-IM under15dB SNR, **e** is the constellation of received signals in WHT-OFDM-IM under 15 dB SNR, and **f** is the constellation of received signals in Double WHT-OFDM-IM under 15 dB SNR

Moreover, the BER and SER performance versus SNR of Double WHT-OFDM-IM, WHT-OFDM-IM and OFDM-IM are investigated in Fig. 5. It compares the performance of Double WHT-OFDM-IM, WHT-OFDM-IM, and OFDM-IM, and reveals that Double WHT-OFDM-IM achieves optimal transmission performance. It considerably reduces the SER and BER compared to OFDM-IM, and there is an approximately 23 dB SNR penalty when the BER is 3.8×10^{-3} compared to conventional OFDM-IM. These findings are consistent with the theoretical properties described in [16], which indicate that not only can Double WHT improve the transmission performance of OFDM, but also OFDM-IM.

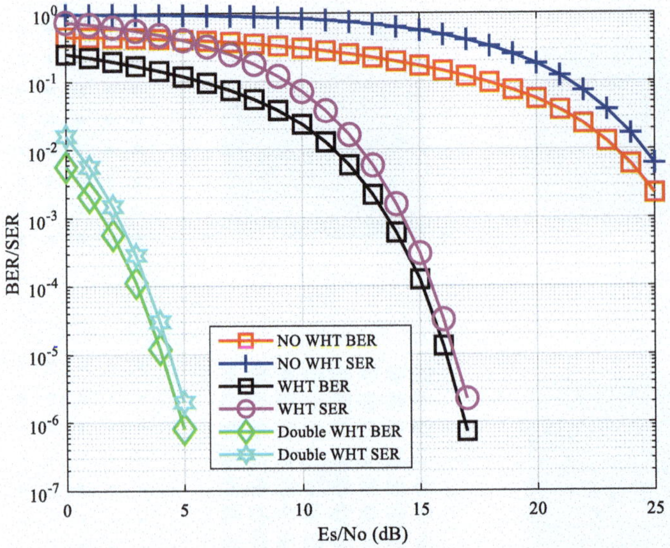

Fig. 5 Simulation of the BER and SER between OFDM-IM, WHT-OFDM-IM and double WHT-OFDM-IM

5 Conclusion

In conclusion, this paper proposes the use of Double WHT-OFDM-IM systems to improve transmission performance and combat the high PAPR of OFDM-IM signals. By incorporating WHT and optimal parameters, the Double WHT-OFDM-IM systems demonstrate high efficiency and PAPR-friendliness through adaptive and dynamic selection between WHT and Double WHT. Simulation results show that the Double WHT-OFDM-IM systems effectively address the problem of distortion caused by OFDM-IM signals passing through non-linear devices with high PAPR, and significantly reduce BER and SER compared to conventional OFDM-IM. The potential of Double WHT-OFDM-IM to improve system transmission efficiency and decrease the cost of HPA make it a highly attractive candidate for the next generation of wireless communication systems. Future research can further explore how to optimize and implement the system in more complex communication environments.

References

1. Korrai P, Lagunas E, Sharma SK, Chatzinotas S, Bandi A, Ottersten B (2020) A RAN resource slicing mechanism for multiplexing of eMBB and URLLC services in OFDMA based 5G wireless networks. IEEE Access 8:45674–45688
2. Lai Y-T, Ciou Y-R, Wu J-M (2018) Index modulation multiple access. In: 2018 IEEE 29th annual international symposium on personal, indoor and mobile radio communications (PIMRC) (2018)
3. Abu-alhiga R, Haas H (2009) Subcarrier-index modulation OFDM. In: 2009 IEEE 20th international symposium on personal, indoor and mobile radio communications, pp 177–181 (2009). https://doi.org/10.1109/PIMRC.2009.5449882
4. Basar E, Aygolu U, Panayirci E (2013) Orthogonal frequency division multiplexing with index modulation. IEEE Trans Signal Process 61(22):5536–5549
5. Van Luong T, Ko Y, Vien NA, Nguyen DHN, Matthaiou M (2019) Deep learning-based detector for OFDM-IM. IEEE Wirel Commun Lett 8(4):1159–1162
6. Mahmuda S, Haque TN, Ahmed F (2012) Adaptive companding as a PAPR reduction technique of an OFDM signal. J Commun 7(11)
7. Lv X, Wan Y (2014) A new weighted tone reservation method for PAPR reduction in OFDM systems. J Commun
8. Wang M. School of electrical engineering and information. South-West Petroleum University; Xiao B (2012) A PAPR reduction method based on differential evolution. J Commun 7(11)
9. Wang H. College of information science & technology. Hainan University, Haikou, China; Cui Y, DuW, Xu L (2016) Hybrid PAPR reduction scheme with partial transmit sequence and tone reservation for FBMC/OQAM. In: J Commun
10. XiaY. School of physics and electronic information. Luoyang Normal University, Luoyang 471022, China; Zhang J (2016) PAPR reduction for OFDM/OQAM signals using offset-symbols joint SLM method. J Commun
11. Kiambi S. School of engineering. University of Nairobi; Kenya S, Mwangi E, Kamucha G (2021) PAPR reduction in MIMO-OFDM systems using low-complexity additive signal mixing. J Commun 468–478
12. Zheng P, Huang J (2018) Efficient encrypted images filtering and transform coding with Walsh-Hadamard transform and parallelization. IEEE Trans Image Process 27(5):2541–2556
13. Beauchamp K (1984) Applications of Walsh and related functions, with an introduction to sequency theory. Academic, New York, NY, USA
14. Dlugaszewski Z, Wesolowski K (2000) WHT/OFDM—an improved OFDM transmission method for selective fading channels. In: Proceedings of the IEEE Benelux chapter vehicular technology communication symposium on communication and vehicular technology (SCVT), pp 144–149 (2000)
15. Wu Y, Ho CK, Sun S (2003) On some properties of Walsh-Hadamard transformed OFDM. In: Proceedings IEEE 56th vehicular technology conference
16. Usman MR, Khan A, Usman MA, Shin SY (2018) Joint non-orthogonal multiple access (NOMA) & Walsh-Hadamard transform: enhancing the receiver performance. China Commun 15(9):160–177

Connectivity Probability Analysis for Freeway Vehicle Scenarios in Vehicular Networks

Weiyi Ni, Qiuyu Zhang, and Hailin Xiao

Abstract The vehicular network's connectivity probability can provide theoretical guidance for both reducing traffic accidents and enhancing traffic efficiency. Most of the previous studies are focused on connectivity probability in ideal freeway scenarios, without considering the impact of entrances/exits. In this paper, a dynamic connectivity analytical model for freeway scenarios with entrances/exits is proposed. In the proposed model, some parameters, including the vehicle flow, the communication radius, the vehicle speed, and the vehicles' travel probability through the entrances/exits, are dynamically set and analyzed in the practical freeway scenarios. The connectivity probability analysis is verified through simulation. Numerical results showed that the above-mentioned parameters, except for the vehicle speed, are closely correlated with the connectivity probability.

Keywords Vehicular ad hoc networks (VANET) · Vehicle-to-vehicle (V2V) communications · Connectivity probability

1 Introduction

The connectivity for vehicular communication is very important for transmitting warning messages generated by a vehicle that detects an accident to the following vehicles [1]. In vehicular networks, vehicular communication can usually be categorized into two modes: vehicle-to-infrastructure (V2I) communications and vehicle-to-vehicle (V2V) communications [2]. Due to V2V communications with low

W. Ni
Changchun University of Science and Technology, Changchun 130022, China

Q. Zhang
Guilin University of Electronic Technology, Guilin 541004, China

H. Xiao (✉)
Hubei University, Wuhan 430062, China
e-mail: xhl_xiaohailin@163.com

M. Ma (ed.), *Proceedings of the 12th International Conference on Communications, Circuits, and Systems*, Lecture Notes in Electrical Engineering 1193,
https://doi.org/10.1007/978-981-97-2636-3_2

latency to transmit traffic information or warning messages, V2V communications is regarded as one of the best alternatives for avoiding high delay introduced by V2I communications [3].

Connectivity probability is a pivotal metric measure of vehicular networks. Especially in a freeway scenario, the highly dynamic characteristic has led to data delivery delay and reduced the vehicular networks performance. Therefore, connectivity probability was widely studied for freeway scenarios [2–8]. Note that most studies have focused on ideal freeway scenarios without entrances/exits. In practice, vehicles may enter, leave, or stay at freeway different locations through entrances/exits, and thus making the networks connectivity highly dynamic. In [9], only an entrance/exit on the freeway has been presented (because these entrances/exits are far away from each other), in which scenarios the entrances/exits can be regarded as independent of each other. Therefore, the research of freeway scenarios with an entrance/exit has representative. In this paper, we present a dynamic connectivity analytical model in freeway scenarios with an entrance/exit. In the proposed model, some parameters are dynamically set and analyzed in the actual freeway scenarios, including vehicle speed, vehicle flow, communication radius, and vehicles travel probability.

2 System Model

We consider a V2V communication scenarios in vehicular Ad-hoc network (VANET). In the proposed system model, each vehicle is equipped with a single antenna. The instantaneous channel gains are assumed the quasi-static Rayleigh distribution. This is reasonable for VANET V2V communications, and some experimental results also demonstrated that the applicability of Rayleigh channel models in practice [7]. Furthermore, in previous works, a perfect channel state information (CSI) is assumed at the transmitters/receivers, where the assumption can be utilized in practice [10]. In freeway scenarios, the entrance/exit on the freeway divides the freeway into two segments, with each vehicle travelling through the entrance/exit with a given probability.

Here, vehicle mobility and highly dynamic traffic environment constitute the main features of vehicular networks. Considering the characters of vehicular networks, the vehicle flow λ are described as Poisson distribution, the vehicle speed V follows a Gaussian distribution and communication radius R is a random variable [6].

The typical actual freeway scenarios are shown in Fig. 1. The entrance/exit is located at r ($r \in [0, d]$), and the freeway is divided into two segments, i.e., $[0, r]$ and $[r, d]$. The vehicles arrive in segment $[0, r]$ will either remain moving in segment $[r, d]$ or leave from the segment $[0, d]$. Furthermore, the vehicles include two parts: the vehicles coming from segment $[0, r]$ and keeping to travel in segment $[r, d]$; on the other hand, new arriving vehicles from the entrance/exit. Therefore, the vehicle flow in segment $[r, d]$ can be written as

$$\lambda_2 = \alpha\lambda_1 + \lambda_N \tag{1}$$

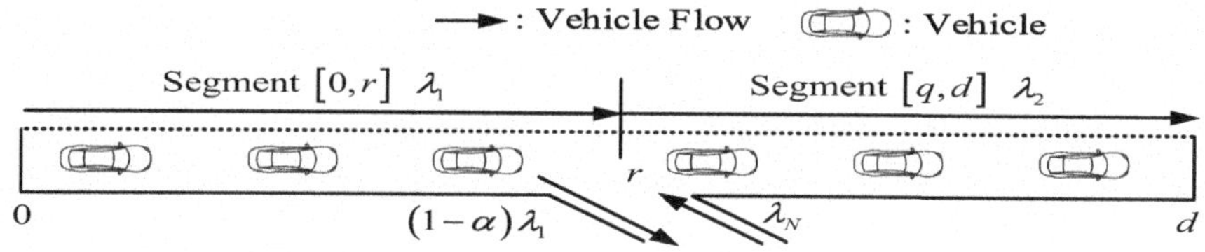

Fig. 1 A simplified model of the scenarios

where α denotes the travel probability of vehicles through the entrance/exit, λ_1 denotes the vehicle flow in segment $[0, r]$, and λ_N denotes the new arriving vehicle flow in segment $[r, d]$.

According to [6], we obtain the vehicle densities ρ_1 and ρ_2 in segments $[0, r]$ and $[r, d]$, respectively.

$$\rho_1 = \frac{\frac{2\lambda_1}{\sqrt{2\pi}\sigma}}{erf\left(\frac{v_{\max}-\mu}{\sigma\sqrt{2}}\right) - erf\left(\frac{v_{\min}-\mu}{\sigma\sqrt{2}}\right)} \int_{v_{\min}}^{v_{\max}} \frac{1}{v} e^{\frac{-(v-\mu)}{2\sigma^2}} dv \tag{2}$$

where $v_{\max} = \mu + 3\sigma$ denotes the biggest vehicle speed, $v_{\min} = \mu - 3\sigma$ is the smallest vehicle speed, μ is the vehicle speed, and σ denotes the vehicle speed standard deviation.

3 Analysis of Connectivity Probability

Now, we will derive the connectivity probability of all vehicles on the freeway. To ensure transmission information in time from a source vehicle to the other vehicles, Any vehicles can directly communicate or via multiple hops communicate between intermediate vehicles [6].

Figure 2 shows two cases where the entrance/exit is lain in $[0, R]$ and $[R, 2R]$. And we will give the analysis processes of case 1 and case 2. Readers can also analyze other cases on their own according to our analytical method.

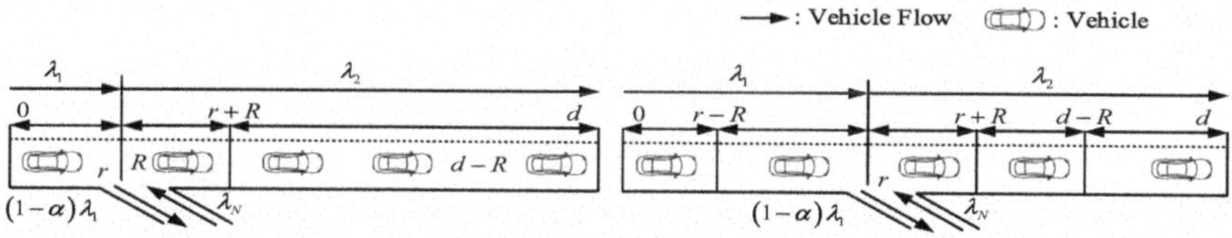

Fig. 2 A general connectivity scenario includes two cases, **a** case 1, **b** case 2

Case 1: $r \in [0, R]$.

As shown in Fig. 2a, the location of the entrance/exit is lain in $[0, R]$, i.e., $r \in [0, R]$. Note that the vehicles can be connected directly if the length of a segment is less than R, otherwise, they will be connected via multi-hop. To accurate analysis connectivity probability, the segment $[0, d]$ is divided into three sub-segments: $[0, r]$, $[r, r + R]$ and $[r + R, d]$. According to vehicles distribution in the different segments, Case 1 can be divided into the following three sub-cases:

Case 1.1: there are vehicles in each sub-segment;

In this case, the connectivity probability can be obtained the segment $[0, d]$ as

$$\begin{aligned} P_{11} = & P_{V[0,r]}(\rho_1, r) P_{W[r,d]}(\rho_2, d - r) P_{Z[0,r+R]}(\rho_1, \rho_2, r, R) P_{H[0,r]}(\rho_1, r) \\ & \times P_{H[r,r+R]}(\rho_2, R) P_{H[r+R,d]}(\rho_2, d - r - R) \end{aligned} \quad (3)$$

where $P_{V[0,r]}(\rho_1, r)$ is the probability that the vehicles in the segment $[0, r]$ can communicate directly, $P_{W[r,d]}(\rho_2, d - r)$ denotes the probability that the vehicles in the segment $[r, d]$ can communicate via multiple hops between intermediate vehicles, $P_{Z[0,r+R]}(\rho_1, \rho_2, r, R)$ is the probability that the vehicles in the segment $[0, r]$ can communicate with the vehicles in the segment $[r, r + R]$, and $P_{H[0,r]}(\rho_1, r)$ is the probability there exist vehicles in the segment $[0, r]$.

$P_{V[0,r]}(\rho_1, r)$ and $P_{W[r,d]}(\rho_2, d - r)$ are respectively calculated by following [6, 7]

$$P_{V[0,r]}(\rho_1, r) = 1 + \frac{\gamma_{th} P_{noise} r^2}{\beta p_t} e^{-\rho_1 r} - \frac{2\gamma_{th} P_{noise}}{\rho_1^2 \beta p_t} + \frac{2\gamma_{th} P_{noise} r}{\rho_1 \beta p_t} e^{-\rho_1 r} + \frac{2\gamma_{th} P_{noise}}{\rho_1^2 \beta p_t} e^{-\rho_1 r} \quad (4)$$

$$P_{W[r,d]}(\rho_2, d - r) = \left(P_{V[r,d]}(\rho_2, d - r)\right)^{(\rho_2, d - r) - 1} \quad (5)$$

where γth denotes the SNR threshold, P_{noise} is additive noise power, β is a path- loss parameter, and p_t denotes the transmit power.

According to [3], we have

$$P_{Z[0,r+R]}(\rho_1, \rho_2, r, R) = 1 - \sum_{k=0}^{\infty} P_{N[r,r+R]}(\rho_2, R, k) (P_{S[0,r+R]}(\rho_1, \rho_2, r, R))^k \quad (6)$$

where $P_{N[r,r+R]}(\rho_2, R, k)$ denotes probability that there occur k vehicles in the segment $[r, r + R]$ and $P_{S[0,r+R]}(\rho_1, \rho_2, r, R)$ denotes the probability that vehicles in segment $[0, r]$ cannot communicate with any vehicles in segment $[r, r + R]$.

$P_{S[0,r+R]}(\rho_1, \rho_1, r, R)$ can be given as (7) (shown at the top of next page), where $x_{[0,r]}$ is the distance between two neighboring vehicles in the segment $[0, r]$ and $f(\rho_1, x_{[0,r]}) = \rho_1 e^{-\rho_1 x_{[0,r]}}$ denotes the probability density function (PDF) of $x_{[0,r]}$.

$$P_{S[0,r+R]}(\rho_1,\rho_2,r,R)=\frac{\sum_{k=1}^{\infty}P_{N[0,r]}(\rho_1,r,k)\Big(1-\int_0^R\int_0^r f(\rho_1,x_{[0,r]})f(\rho_2,x_{[r,r+R]})dx_{[0,r]}dx_{[r,r+R]}\Big)^k}{1-P_{\overline{H}[0,r]}(\rho_1,r)} \tag{7}$$

Usually, vehicles' flow can be described as Poisson process [11], and thus we can obtain the following results

$$P_{N[r,r+R]}(\rho_2,R,k)=\frac{(\rho_2 R)^k e^{-\rho_2 R}}{k!}, k\geq 0 \tag{8}$$

$$P_{H[0,r]}(\rho_1,r)=1-P_{H[0,r]}(\rho_1,r)=P_{N[0,r]}(\rho_1,R,0) \tag{9}$$

By more concise descriptions, the remainder of Case 1 and the analysis of Case 2 are shown in Table 1. Let $H_{[0,r-R]}$ denote there exist vehicles in the segment $[0, r-R]$, and $\overline{H}_{[0,r-R]}$ denote there doesn't exist vehicle in the segment $[0, r-R]$.

4 Numerical Simulation Results

Numerical results are verified the connectivity probability performance for the parameters λ, R, μ and α. Moreover, some parameters are found (reference in [6, 9]) unless stated otherwise. Note that for each scenario in the proposed simulation, 1000 iterations are performed.

Figure 3 depicts the vehicle flow λ versus the connectivity probability of Case 1. As expected, the higher λ_1 and λ_N imply a higher connectivity probability, in the case the influence of λ_1 on the connectivity probability is relatively high. It demonstrates that the numerical results are consistent with the analytical results.

Figure 4 evaluates connectivity probability of the different vehicle speeds μ as a function of communication radius R. The vehicular networks connectivity probability will be improved as R increases. It can be concluded that the communication radius effects on connectivity probability greatly. However, when R is greater than 150 km, the effect of R on the connectivity probability becomes negligible. In this case, the connectivity probability will be independent of R. On the other hand, the connectivity probability decreases as μ increases, because the distance of two neighbor vehicles leads to connectivity probability decrease.

Figure 5 depicts the connectivity probability versus α for Case 1 and Case 2, respectively. Numerical results indicate that increase α will improve the connectivity probability, because α increases as the vehicle density ρ_2 increases. It demonstrates that the Case 2's connectivity probability is smaller than that of Case 1 (i.e., $P_1 > P_2$). Therefore, when the location of entrance/exit is lain at the end of the freeway, the vehicular networks have better connectivity.

Table 1 The connectivity probability analysis of case 1.2, case 1.3 and case 2

Case 1.2: There are vehicles in two sub-segments	
1. $\overline{H}_{[0,r]}$	$P_{121} = P_{W[r,d]}(\rho_2, d-r)P_{\overline{H}[0,r]}(\rho_1, r)P_{H[r,r+R]}(\rho_2, R)P_{H[r+R,d]}(\rho_2, d-r-R)$
2. $\overline{H}_{[r+R,d]}$	$P_{122} =$ $P_{V[0,r]}(\rho_1, r)P_{V[r,r+R]}(\rho_2, R)P_{Z[0,r+R]}(\rho_1, \rho_2, r, R)P_{H[0,r]}(\rho_1, r)P_{H[r,r+R]}(\rho_2, R)P_{\bar{H}[r,R,d]}(\rho_2, d-r-R)$
Case 1.3: There are vehicles in one sub-segment	
1. $H_{[0,r]}$	$P_{131} = P_{V[0,r]}(\rho_1, r)P_{H[0,r]}(\rho_1, r)P_{\overline{H}[r,d]}(\rho_2, d-r)$
2. $H_{[r,r+R]}$	$P_{132} = P_{V[r,r+R]}(\rho_2, R)P_{\overline{H}[0,r]}(\rho_1, r)P_{H[r,r+R]}(\rho_2, R)P_{\overline{H}[r+R,d]}(\rho_2, d-r-R)$
3. $H_{[r+R,d]}$	$P_{133} = P_{W[r+R,d]}(\rho_2, d-r-R)P_{\overline{H}[0,r]}(\rho_1, r)P_{\overline{H}[r,r+R]}(\rho_2, R)P_{H[r+R,d]}(\rho_2, d-r-R)$
Case 1: $P_1 = P_{11} + \sum_{i=1}^{2} P_{12i} + \sum_{i=1}^{3} P_{13i}$	
Case 2: $r \in [R, 2R]$	
Case 2.1: There are vehicles in each sub-segment	
$P_{21} = P_{W[0,r]}(\rho_1, r)P_{W[r,d]}(\rho_2, d-r)P_{Z[r-R,r+R]}(\rho_1, \rho_2, R, R)$ $P_{H[0,r-R]}(\rho_1, r-R)\rho_{H[r-R,r]}(\rho_1, R)P_{H[r,r+R]}(\rho_2, R)P_{H[r+R,d]}(\rho_2, d-r-R)$	
Case 2.2: There are vehicles in three sub-segments	
1. $\overline{H}_{[0,r-R]}$	$P_{221} = P_{V[r-R,r]}(\rho_1, R)P_{W[r,d]}(\rho_2, d-r)P_{Z[r-R,r+R]}(\rho_1, \rho_2, R, R)P_{\overline{H}[0,r-R]}(\rho_1, r-R)$ $\times P_{H[r-R,r]}(\rho_1, R)P_{H[r,r+R]}(\rho_2, R)P_{H[r+R,d]}(\rho_2, d-r-R)$
2. $\overline{H}_{[r+R,d]}$	$P_{222} = P_{W[0,r]}(\rho_1, r)P_{V[r,r+R]}(\rho_2, R)P_{Z[r-R,r+R]}(\rho_1, \rho_2, R, R)P_{H[0,r-R]}(\rho_1, r-R)$ $\times P_{H[r-R,r]}(\rho_1, R)P_{H[r,r+R]}(\rho_2, R)P_{\overline{H}[r+R,d]}(\rho_2, d-r-R)$

(continued)

Table 1 (continued)

Case 2.3: There are vehicles in two sub-segments	
1. $\overline{H}_{[0,r]}$	$P_{231} = P_{W[r,d]}(\rho_2, d-r)P_{\overline{H}[0,r]}(\rho_1, r)P_{H[r,r+R]}(\rho_2, R)P_{H[r+R,d]}(\rho_2, d-r-R)$
2. $\overline{H}_{[0,r-R]}$, $\overline{H}_{[r+R,d]}$	$P_{232} = P_{V[r-R,r]}(\rho_1, R)P_{V[r,r+R]}(\rho_2, R)P_{Z[r-R,r+R]}(\rho_1, \rho_2, R, R)$ $\times P_{\overline{H}[0,r-R]}(\rho_1, r-R)P_{H[r-R,r]}(\rho_1, R)P_{H[r,r+R]}(\rho_2, R)P_{\overline{H}[r+R,d]}(\rho_2, d-r-R)$
3. $\overline{H}_{[r,d]}$	$P_{233} = P_{W[0,r]}(\rho_1, r)P_{H[0,r-R]}(\rho_1, r-R)P_{H[r-R,r]}(\rho_1, R)P_{\overline{H}[r,d]}(\rho_2, d-r)$
Case 2.4: There are vehicles in one sub-segment	
1. $H_{[0,r-R]}$	$P_{241} = P_{V[0,r-R]}(\rho_1, r-R)P_{H[0,r-R]}(\rho_1, r-R)P_{\overline{H}[r-R,r]}(\rho_1, R)P_{\overline{H}[r,d]}(\rho_2, d-r)$
2. $H_{[r-R,r]}$	$P_{242} = P_{V[r-R,r]}(\rho_1, R)P_{\overline{H}[0,r-R]}(\rho_1, r-R)P_{H[r-R,r]}(\rho_1, R)P_{\overline{H}[r,d]}(\rho_2, d-r)$
3. $H_{[r,r+R]}$	$P_{243} = P_{V[r,r+R]}(\rho_2, R)P_{\overline{H}[0,r]}(\rho_1, r)P_{H[r,r+R]}(\rho_2, R)P_{\overline{H}[r+R,d]}(\rho_2, d-r-R)$
4. $H_{[r,r+R]}$	$P_{244} = P_{W[r+R,d]}(\rho_2, d-r-R)P_{\overline{H}[0,r]}(\rho_1, r)P_{\overline{H}[r,r+R]}(\rho_2, R)P_{H[r+R,d]}(\rho_2, d-r-R)$
Case 2: $P_2 = P_{21} + \sum_{i=1}^{2} P_{22i} + \sum_{i=1}^{3} P_{23i} + \sum_{i=1}^{4} P_{24i}$	

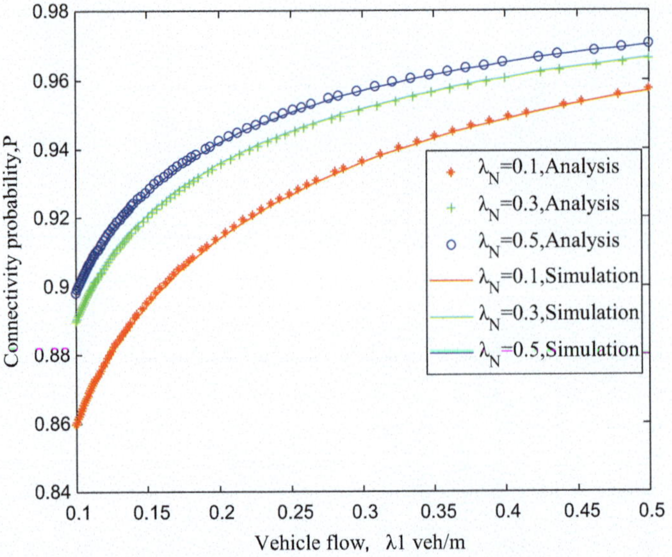

Fig. 3 Connectivity probability versus vehicle flow λ veh/ s ($R = 500$ m, $\mu = 30.6$ m/s, $\alpha = 0.5$)

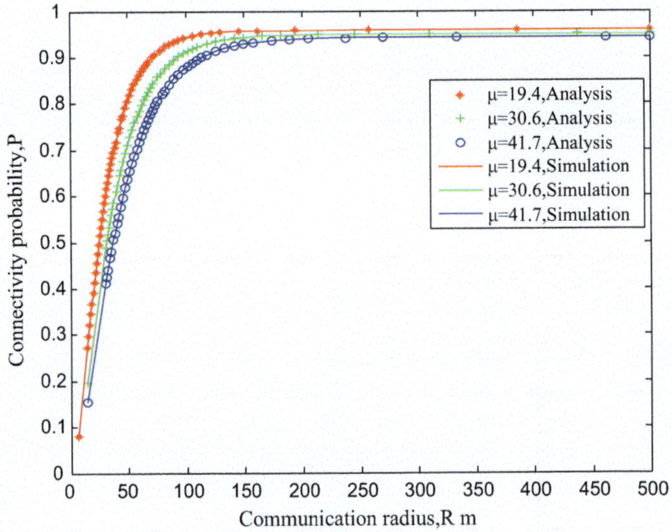

Fig. 4 Connectivity probability versus distance R m ($\lambda_1 = 0.3$ veh/ s, $\lambda_N = 0.3$ veh/ s, $\alpha = 0.5$) for different μ m/s

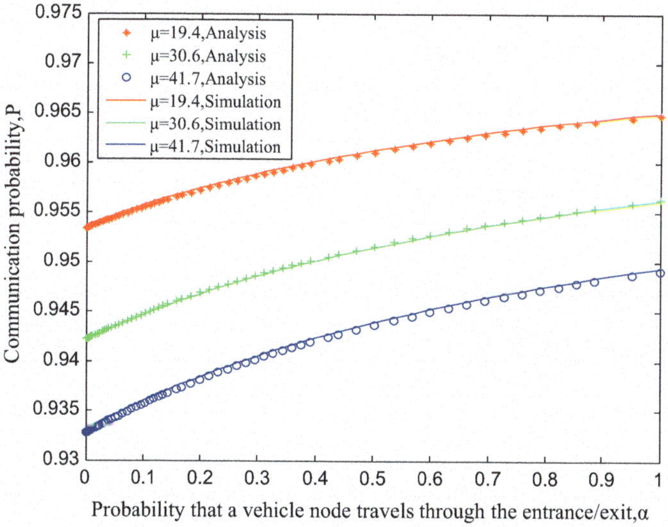

Fig. 5 Connectivity probability versus $\alpha(\lambda_1 = 0.3$ veh/ s, $\lambda_N = 0.3$ veh/ s, $\mu = 30.6$ m/ s, $R = 500)$ for Case 1 and Case 2

5 Conclusion

In this paper, the connectivity probability expressions of vehicular networks have been derived on the freeway with an entrance/exit in the practical scenarios. Some parameters λ, V, R and α for connectivity probability have been analyzed via both numerical and theoretical analysis. It is shown that the numerical results are consistent with the theoretical analysis. Moreover, both increasing parameters (λ, R and α) and decreasing μ can substantially improve the connectivity probability. More importantly, the obtained results are capable of providing some theoretical guidance to reduce the traffic accidents and enhance the traffic efficiency.

References

1. Xiao H, Zhang Q, Ouyang S, Chronopoulos AT (2021) Connectivity probability analysis for VANET freeway traffic using a cell transmission model. IEEE Syst J 15(2):1815–1824
2. Kwon S, Kim Y, Shroff NB (2016) Analysis of connectivity and capacity in 1-D vehicle-to-vehicle networks. IEEE Trans Wirel Commun 15(12):8182–8194
3. Gao H, Liu C, Li Y, Yang X (2021) V2VR: reliable hybrid-network-oriented V2V data transmission and routing considering RSUs and connectivity probability. IEEE Trans Intell Transp Sys 22(6):3533–3546
4. Ahmed RE (2011) A Fault-tolerant routing protocol for mobile ad hoc networks. J Adv Inf Technol 2(2):128–132
5. Abdelghaffar HM, Elouni M, Bichiou Y, Rakha HA (2020) Development of a connected vehicle dynamic freeway variable speed controller. IEEE Access 8:99219–99226

6. Choi S, Choi J-G, Bahk S (2020) Mobility-aware analysis of millimeter wave communication systems with blockages. IEEE Trans Veh Tech 69(6):5901–5912
7. Oubbati OS, Atiquzzaman M, Baz A, Alhakami H, Ben-Othman J (2021) Dispatch of UAVs for urban vehicular networks: a deep reinforcement learning approach. IEEE Trans Veh Tech 70(12):13174–13189
8. Abuelenin SM, Elaraby S (2022) A generalized framework for connectivity analysis in vehicle-to-vehicle communications. IEEE Trans Intell Transp Syst 23(6):5894–5898
9. Wang Y, Zheng J (2018) Connectivity analysis of a highway with one entry/exit and multiple roadside units. IEEE Trans Veh Tech 67(12):11705–11718
10. Singh R, Saluja D, Kumar S (2020) Reliability improvement in clustering-based vehicular ad-hoc network. IEEE Commun Lett 24(6):1351–1355
11. Wang Y, Zheng J (2016) A connectivity analytical model for a highway with an entrance/exit in vehicular ad hoc networks. In: Proceedings of the IEEE ICC, Kuala Lumpur, Malaysia, pp 1–6

Hybrid Control of Channel Selection and Transmission Power for Low Power Wireless Sensor Network

Chung-Wen Hung and Tzu-Hui Lee

Abstract A hybrid low power control of channel selection and transmission power (TP) which is suitable for Sub-1 GHz Wireless Sensor Network (WSN) is proposed in this paper. Internet of thing (IoT) is popular, and WSN is getting even much hotter due to build convenience and low cost. A large number of devices which share the Sub-1 GHz Industrial Scientific Medical Band (ISM Band) together may cause channel contention. Therefore, the way to maintain the quality of communication with lower power consumption became an important issue. The technique of channel estimation is adopted to select the best channel for communication. The Signal-to-Noise Ratio (SNR) is chosen as the reference to present the state of environment and propagation path. By determining the SNR value, the quality of communication can be detected and the TP or channel will be adjusted when errors occur. The experimental results show that under the proposed hybrid control node the Packet Error Rate (PER) is efficiently reduced due to the choice of stable channel compared with the control group and the TP-controlled node. Moreover, the hybrid control node can ensure the similar quality of communication with better current consumption contrasts with channel-controlled node.

Keywords Wireless sensor network (WSN) · Low-rate wireless personal area network (LR-WPAN) · Channel estimation · Transmission power control (TPC) · Power consumption

---

C.-W. Hung (✉) · T.-H. Lee
Department of Electrical Engineering, National Yunlin University of Science and Technology, Yunlin County, Douliu, Taiwan
e-mail: wenhung@yuntech.edu.tw

T.-H. Lee
e-mail: m11012038@yuntech.edu.tw

M. Ma (ed.), *Proceedings of the 12th International Conference on Communications, Circuits, and Systems*, Lecture Notes in Electrical Engineering 1193,
https://doi.org/10.1007/978-981-97-2636-3_3

1 Introduction

The development of IoT becomes more popular owing to the advancement of wireless communication. Moreover, WSN is used widely in monitoring due to the advantages of easily implemented and lower cost [1–3]. Nowadays, most of IoT devices communicate in the unlicensed ISM radio band. However, radio band which is the medium of wireless communication is limit. Although each device can access the limited spectrum in the same time with Media Access Control (MAC) protocols, the link quality may still be affected by the interference between devices. Therefore, how to ensure the smoothness of channel and maintain a stable commutation become an important topic [4]. The method of channel estimation and selection which can hop to good channels by measuring channel-related parameters is considered as a solution.

A lot of wireless network technique have defined the standard for channel hopping. For example, Bluetooth develops the technique of Frequency-Hopping Spread Spectrum (FHSS) to hop to different channel randomly [5]. It can not only improve the stability of communication by avoiding the channel contention with Wi-Fi packets which also run in the 2.4 GHz band, but also enhance the security. In IEEE 802.15.4e, the specification defined for Low-Rate Wireless Personal Area Network (LR-WPAN), Time Slotted Channel Hopping (TSCH) protocol developed based on Time Division Multiple Access (TDMA) structure is proposed [6]. It is similar with FHSS and it can select a channel by pseudorandom method to against the interference and multipath fading. Nevertheless, we can't assure that the channel selected by random method is stable enough to transmit signal. As a result, a best channel selection method for Sub-1 GHz is proposed in this paper. By measuring the SNR, the state of channels can be determined and a stable one can be chosen to continue the transmission.

In addition to communication quality, power saving is also an important issue in WSN field. Because WSN system is often used for sensing and monitoring, the location of implementing the system may be in the remote areas that are difficult to arrive. Most WSN nodes use batteries as the power source, making them need to operate with low power to prolong the battery life. After picking a good channel, the system proposed in this paper will adjust the transmission power of nodes so that the nodes can transmit data without consuming much power and can maintain the communication quality at the same time. Currently, many studies have proposed different control methods for TP and there are also some researches about low-power system that simultaneously controls TP and Data Rate (DR) [7]. However, there are few algorithms combining channel selection and Transmission Power Control (TPC) have been proposed. Although there are still a few methods that control channel and TP simultaneously, such as the algorithm designed for Wireless Body Area Networks (WBAN) [8]. Nevertheless, the control strategy adopted for WBAN and WSN are different. Since the implement range of WSN is wider than WBAN, nodes of WSN may be affected by the environment and other devices worked in the same spectrum, resulting in communication failure. Therefore, the method proposed in this paper is expected to propose an effective hybrid control scheme of channel and TP for Sub-1 GHz WSN.

As the main issues of this paper, the related research and discussion of channel and TP are introduced in the second and third chapters respectively. Moreover, the research methods are explained in the fourth chapter. The experimental results will be analyzed in the fifth chapter and finally the sixth chapter will give the conclusion of research.

2 Channel Estimation and Selection

2.1 Background

IEEE 802.15.4 has published a frequency hopping standard defined for LR-WPAN devices. This protocol selects the desired channel from 16 specific channels set based on the calculation of the Absolute Sequence Number (ASN) of the network time, channel offset and pseudo-random frequency hopping sequence. However, if the selected channel hasn't measured and isn't good enough compared with currently channel, it may lead to worse results and cause error. Therefore, many studies have proposed the improved methods for TSCH.

The methods of [9, 10] use Link Quality Indicator (LQI) and Received Signal Strength Indication (RSSI) as reference respectively and compare with the self-setting thresholds for determining the link quality and reflecting the stability of connection. Moreover, both methods have a strategy of regularly updating the channel state and sorting to cope with the time variant characteristics of the communication environment. In addition, to against the interference sources operating in a certain frequency band, many algorithms such as [11–13] have introduced the concept of blacklist to help nodes make decisions of channel hopping to avoid unstable frequency bands. The Adaptive Channel Selection (ACS) algorithm indicates the limitations of using a single parameter as a standard parameter [14], therefore, decision-making method that considers multiple parameters such as RSSI and Packet Reception Rate (PRR) is proposed. The Power Control and Blacklisting (PCBL) algorithm forms a blacklist by traversing the Link Quality Estimator (LQE) under different combinations of channel and TP [15]. However, using this method requires additional handshaking that will increase energy consumption. To sum up, we list several conditions that need to be concerned when designing a channel selection algorithm.

- Reference parameter: A suitable parameter need to reflect the situation of transmission channel immediately and can predict the possible occurrence of errors. What's more, the process of measuring the parameter should consume less power.
- Complexity of calculation: The method to processes the reference parameters and to generates the threshold for comparison should be simple enough in order not to put a strain on system and maintain the immediacy of communication.

- Property of time-variant: The above studies have illustrated the importance of regularly updating the available channels. Therefore, we have to design a suitable algorithm that can adjust the current decision adaptively based on the feedback LQEs.

2.2 *Preparation for Algorithm Design*

In order to observe the relationship between the characteristics of channels and communication parameters, an experiment which is designed to analyze the ability of parameters to reflect the link quality by collecting the LQE when communicating with sensing nodes. As shown in Fig. 1, the experiment results show that the interference level of each channel is totally different at the same time, which verifies the necessity of channel hopping during communication. It can also be seen from the graph that the generation of PER is related to changes of RSSI and Noise Floor (NF). However, not every poor NF value will cause errors. For instance, the channel of 875 MHz shown in the picture has worse NF level compared to other channels while its RSSI perform well. Obviously, only consider a single parameter as the reference may let the algorithm misjudges and takes unnecessary control actions which could lead to unstable communication and additional power consumption. Compared with the aforementioned several algorithms which only consider a single parameter and may not be able to reflect the current environmental state completely, the algorithm proposed in this paper decides to use SNR as a reference parameter. By measure the environmental noise of the node and the signal strength of packet in the propagation path simultaneously, the quality of communication can be simulated and a more stable control result can be achieved. The measurement of NF can be obtained by scanning the environment before receiving the packet when the node turns on the RF and prepares to receive the message. It does not require redundant handshaking or additional current consumption, which conform to the goal of this research.

3 Transmission Power Control

The topic of TPC has been discussed in several studies. [16] generates the linear control model between RSSI and TP according to historical data and keep updating the parameters of model through the feedback LQEs of current information during the transmission process. The experimental results prove that when the link quality decrease, the Adaptive Transmission Power Control (ATPC) method which adjust TP value adaptively can control the PRR within the target range of the paper while the node with fixed TP accumulates a large number of transmission errors. However, since TP value is a value calculated by exponential operation, it cannot be completely described by linear relationship. Therefore, the algorithm such as Practical Transmission Power Control (PTPC) is proposed [17]. PTPC uses the Proportional- Integral

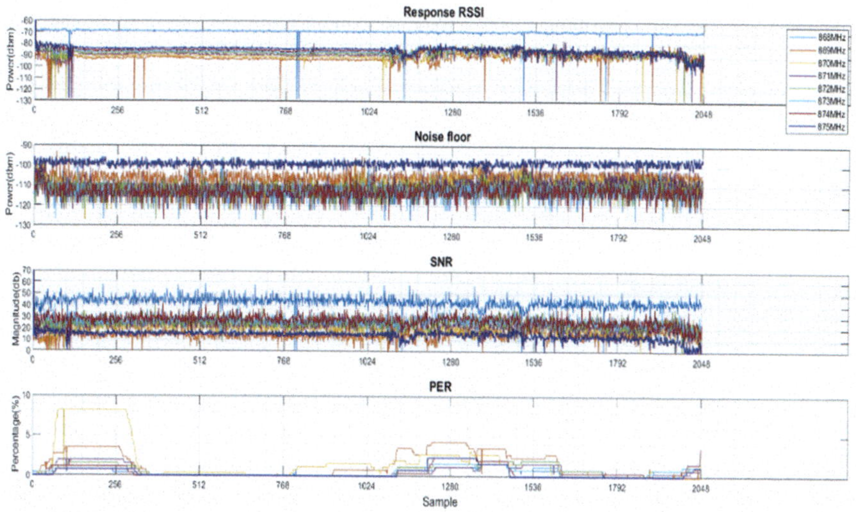

Fig. 1 LQEs of sensing node

(PI) controller which is also a kind of linear controller and takes PRR as the input parameter to calculates the corresponding TP value. Besides, an Anti-Windup control framework is implemented to solve the nonlinear relationship between PRR and TP. The Adaptive and Robust Topology (ART) algorithm considers the continuity of communication in time [18]. It introduces the concept of "Window", using the trend of a specific number of PER as the basis for TP control. LQI and RSSI are used as the main parameters in RSSI/LQI-based Transmission Power Control (RL-TPC) algorithm to judge the current connection quality [19]. The RL-TPC method can determine whether the sensor node is facing the signal weakness or external interferences according to the status of parameters and maintain stable communication by control TP. Authors focus on the process design of handshaking to reach the TP control in the research of WSN application for water level telemetry [20]. After transmitting with maximum TP when communication start, the algorithm keeps adjusting TP according to the response RSSI value by the proposed handshaking mechanism.

All studies above use the current LQE values as feedbacks of the system to generate a suitable model for current environment and resist the interference happened randomly during the transmission process. In addition, the survey mentioned that the correlation between LQI and PRR is stronger compared with the relation of RSSI and PRR, which makes LQI a better indicator to present the link quality [21]. However, when the connection start to decline, the variance of LQI is more significant so that single LQI value is not good enough to evaluate the connection status. It is also mentioned in the research that SNR is a better LQE compared with RSSI owing to its ability to distinguish the quality of the environment. Because SNR is the difference between the RSSI and NF, which means it can describe the state of the link itself and the environment where the node is located at the same time. Hence, the accuracy of PER estimation can be effectively improved.

Owing to the above reasons, SNR is selected as the reference parameter to decide whether the TP value should increases or decreases, too. As mentioned in Chap. 2, SNR of bridge and the sensor nodes are used to consider whether to change the frequency or not. If the SNR at both sides drops at the same time, it means the current environment and the propagation channel are not clear enough, which resulting in the interference on uplink and downlink. However, if SNR of bridge is poor while the SNR of sensor node is not, the system can try to enhance the signal strength to resist possible noise interference by increasing the TP of the sensor node.

4 Experiment Setting

4.1 System Implementation

The system uses the CC1310 [22] developed by Texas Instruments as a communication node to construct a network. It operates in the Sub-1 GHz and uses B-FSK digital modulation technology. The network adopts star topology as architecture of system to establish point-to-point communication between the bridge and each sensor node. In the MAC layer, the technique of TDMA is used. As shown in Fig. 2, TDMA allocates a fixed duration and non-overlapping time slot for each node so that the node can only turn on the RF to communicate with the bridge when its time slot is polled. Therefore, this method can not only avoid channel contention between nodes, but also ensure that sensing nodes can save more power and extend the battery life by turning off RF during other time. In order to verify the feasibility and effectiveness of the system, the proposed system is deployed at different location in building for long-term testing. By comparing the results from different environments can confirm the anti-interference ability of the algorithm and summarize the applicable situations of this algorithm.

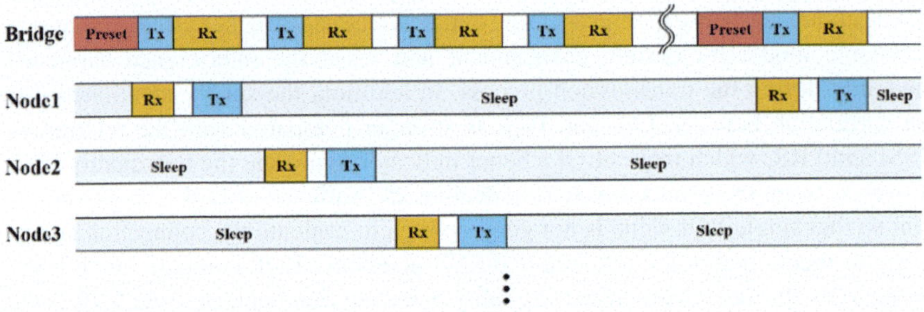

Fig. 2 The structure of TDMA

4.2 Algorithm of Hybrid Control

As mentioned in Chaps. 2 and 3, SNR value of both side are selected as the parameter to detect the environment. SNR of node is calculated by require RSSI, the signal strong of propagation path from bridge to node, and NF of node while the SNR of bridge can be obtained by calculating response RSSI which characterize the signal strong of path from node to bridge. Because bridge is powered by utility power, it always delivers packet with maximum power which equal to 14 dBm for CC1310. Therefore, the only way to prevent a bad communication state when detecting a worse SNR value from node is to select a better channel and hop to new frequency. On the other hand, there are two methods can implement if SNR of bridge isn't good to maintain link quality, one is changing frequency and another is adjusting TP. Table 1 describes the different condition of link based on SNR from both side, helping to select the corresponding control factors for algorithm.

The flow chart of algorithm is shown in Fig. 3. During every communication process, the bridge will judge the link quality based on the received SNR value. The bridge powered by the utility power keeps transmitting with maximum TP, which is equal to 14 dBm, to send requesting packets to the sensing node. When the SNR received by the sensing node decreases, the bridge cannot resist possible channel interference by increasing the transmission power and the system can only hop to a new channel for a more stable environment. Moreover, the instability of link may be eliminated by increasing the TP of the node first when a lower SNR value is received at the bridge. However, if the node still cannot improve the low SNR problem effectively after transmitting with the maximum power, the current channel is considered as an unstable channel so a frequency hopping mechanism is executed to deal with it.

The mechanism that can decrease TP to reduce power consumption when under a stable community environment is also designed for proposed algorithm. If the SNR value of packets during 16 times continuous transmission process are all larger than the threshold, the current environment is determined to be stable enough and the system will try to reduce the TP slowly until the power is small enough without affecting the link quality.

Table 1 Parameters relation

BridgeSNR	NodeSNR	Affect reason	Control factor
Good	Bad	Interference of downlink, environment noise of sensing node	Channel
Bad	Good	Interference of uplink, environment noise of bridge	TP, channel (When TP = TPmax)
Bad	Bad	Interference and noise of both side	Channel

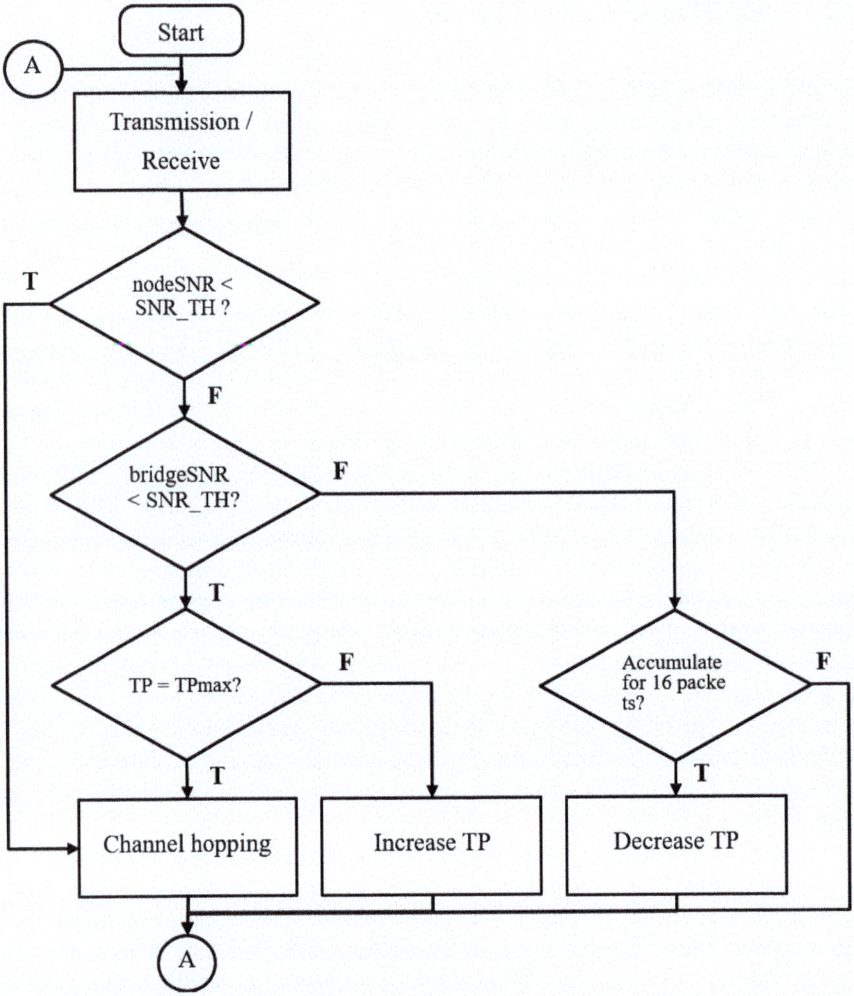

Fig. 3 The flow chart of algorithm proposed by this paper

5 Experiment Result

In order to prove that controlling the communication channel and transmission power at the same time can bring better results to the system, the system also adds the single-element control nodes with channel control and TP control in addition to the sensing nodes that implement hybrid control algorithms and the compared node without any algorithm. The conditions which are used to determine the communication environment and to decide whether to change corresponding parameters are identical for every control node. It helps us to put more attention on the difference between single-element and dual-element control. In addition, the design of algorithm such as

Table 2 Configuration of nodes

Number of node	Transmission power	Algorithm
2, 6, 10, 14	Adjust by algorithm	TP control
4, 8, 12, 16	Adjust by algorithm	Hybrid contol
1	0 dBm	None
3	0 dBm	Channel control
5	2 dBm	None
7	2 dBm	Channel control
9	3 dBm	None
11	3 dBm	Channel control
13	5 dBm	None
15	5 dBm	Channel control

the reduction mechanism of TP control node and the way to select a best channel for channel control node are also adjusted to be consistent with the hybrid control node. It is expected to simulate and compare the effects of using different control items under the same communication conditions, making hybrid control algorithm a better choice for WSN. The experiments is set up at four locations. Because the distance between each site and bridge is different, each test group will use different initial TP at the beginning of communication in order to avoid communication failure caused by the distance. Table 2 is the configuration of the experimental nodes.

This paper organizes the experiment results as the following Table 3 to analysis the effect of each algorithm easily. From the results, almost all nodes with algorithm in each group can obtain better PER values than the controlled group. It proves that nodes which can adjust their own communication parameters adaptively have the ability to against interference. Next we compare the result of hybrid control and the other two algorithms. Location 1 is close to bridge and there are less people passing by. Therefore, the environment is stable. Although the PER result of hybrid control algorithm is relatively high in this place, it can still maintain the communication quality which is similar to other two control nodes with the minimum power. The result is acceptable. In second place where the environment is slightly worse, two single-parameter control groups have better results in terms of PER performance and power consumption respectively. Locations 3 and 4 are adjacent to the office and there are many people walk around the experiment group, so the environment changes greatly and communication failure often happen. From the results, it is clear that under such conditions, frequency hopping may not be able to obtain the good communication quality. What's more, it may even cause more failure due to the unstable environment, increasing the error rate and power consumption. On the contrary, the nodes with transmission power control both obtain a good result. Among them, the hybrid control algorithm at location 4 has the best test results. However, because of the extra power consumption caused by frequency hopping, the power consumption is a little higher when the PER value of hybrid control node is similar to TP control node in location 3. According the experiment result, the conclusion can

Table 3 Result of experiment

Number of nodes	Location	PER (%)	Average current consumption (uA)	Hopping time
1	1	1.307	17.65	
2		1.15	17.67	
3		1.2	17.72	42
4		1.24	17.654	0
5	2	1.721	18.132	
6		1.23	17.74	
7		1.213	17.78	32
8		1.37	17.7	35
9	3	11.90	18.083	
10		2.176	17.82	
11		3.1	18.05	441
12		2.2	17.9	149
13	4	2.334	18.345	
14		1.632	17.73	
15		7.001	18.874	965
16		1.0	17.7	0

be observed as following: the hybrid control algorithm which can adjust channel and TP at the same time has better effect on saving power under the good environment while it has a more significant effect on the suppression of transmission errors when the environment is unstable.

6 Conclusion

The proposed hybrid control algorithm has indeed achieved the preset goal: to transmit with the lowest power consumption while maintaining good communication quality. In the future, the model combined current research and adaptive data rate mechanism is expected to achieve a more precise control. In addition to adding new control factor, trying to combine different nonlinear control methods with current research results of channel is also a solution. For instance, the work of research [23] is to detecting the traffic anomalies in ultra-low powered wireless networks based on the combination of attributes of TSCH and machine learning technology. It may provide a more precise result of prediction of channel status.

Acknowledgements This research is supported by National Science and Technology Council, Taiwan, under contract: NSTC 111-2221-E-150-040, and 111-2622-E-224-013.

References

1. Ullo SL, Sinha GR (2020) Advances in smart environment monitoring systems using IoT and sensors. Sensors 20(11):3113. https://doi.org/10.3390/s20113113
2. Gao S, Dai X, Hang Y, Guo Y, Ji Q (2018) Airborne wireless sensor networks for airplane monitoring system. Wirel Commun Mob Comput 2018:1–18. 10.1155 /2018/6025825
3. Yi WY, Lo KM, Mak T, Leung KS, Leung Y, Meng ML (2015) A survey of wireless sensor network based air pollution monitoring systems. Sensors 15(12):31392–31427. https://doi.org/10.3390/s151229859
4. Hasan A, Khan BM (2022) Coexistence management in wireless networks-a survey. IEEE Access 10:38600–38624. https://doi.org/10.1109/ACCESS.2022.3165223
5. Crow BP, Widjaja I, Kim JG, Sakai PT (1997) IEEE 802.11 wireless local area networks. IEEE Commun Mag 35(9):116–126. https://doi.org/10.1109/35.620533
6. IEEE, 802.15.4. Wireless medium access control and physical layer specifications for low rate wireless personal area networks (LR-WPANs) (2015)
7. Hung CW, Zhang HJ, Hsu WT, Zhuang YD (2020) A low-power WSN protocol with ADR and TP hybrid control. Sensors 20(20):5767. https://doi.org/10.3390/s20205767
8. Di Franco F, Tachtatzis C, Atkinson RC, Tinnirello I, Glover IA (2015) Channel estimation and transmit power control in wireless body area networks. IET Wirel Sens Syst 5(1):11–19. https://doi.org/10.1049/iet-wss.2013.0070
9. Yoon SU, Murawski R, Ekici E, Park S, Mir ZH (2010) Adaptive channel hopping for interference robust wireless sensor networks. In: 2010 IEEE international conference on communications, pp 1–5. https://doi.org/10.1109/ICC.2010.5502780
10. Gu J, Xu Y, Zhu Y, Mei F, Kang L, Liu Y (2020) An improved fast transform algorithm based on adaptive hybrid channel allocation in TSCH networks. EURASIP J Wirel Commun Netw (171). https://doi.org/10.1186/s13638-020-01786-2
11. Fabbri F, Zuniga M, Puccinelli D, Marrón P (2012) On the optimal blacklisting threshold for link selection in wireless sensor networks. In: Proceedings of the wireless sensor networks—European conference (EWSN), pp 147–162. https://doi.org/10.1007/978-3-642-28169-3_10
12. Kotsiou V, Papadopoulos GZ, Zorbas D, Chatzimisios P, Théoleyre F (2019) Blacklisting-based channel hopping approaches in low-power and Lossy networks. IEEE Commun Mag 57(2):48–53. https://doi.org/10.1109/MCOM.2018.1800362
13. Du P, Dr. Roussos G (2012) Adaptive time slotted channel hopping for wireless sensor networks. In: 2012 4th computer science and electronic engineering conference (CEEC), pp 29–34. https://doi.org/10.1109/CEEC.2012.6375374
14. Elsts A, Fafoutis X, Piechocki R, Craddock I (2017) Adaptive channel selection in IEEE 802.15.4 TSCH networks. In: 2017 global internet of things summit (GIoTS), pp 1–6. https://doi.org/10.1109/GIOTS.2017.8016246
15. Son D, Krishnamachari B, Heidemann J (2007)Experimental study of transmission power control and blacklisting based link quality control in wireless sensor networks. Information Sciences Institute Technical Report, ISI-TR-629
16. Lin S, Miao F, Zhang J, Zhou G, Gu L, He T, Stankovic JA, Son S, Pappas GJ (2016) ATPC: adaptive transmission power control for wireless sensor networks. ACM Trans Sens Netw 12(1, 6):1–31. https://doi.org/10.1145/2746342
17. Fu Y, Sha M, Hackmann G, Lu C (2012) Practical control of transmission power for wireless sensor networks. In: 2012 20th IEEE international conference on network protocols (ICNP), pp 1–10. https://doi.org/10.1109/ICNP.2012.6459981
18. Chincoli M, Syed AA, Exarchakos G, Liotta A (2016) Power control in wireless sensor networks with variable interference. Mob Inf Syst 2016(3592581):1–10. https://doi.org/10.1155/2016/3592581
19. Kim S, Kim S, Eom D-S (2013) RSSI/LQI-based transmission power control for body area networks in healthcare environment. IEEE J Biomed Health Inform 17(3):561–571. https://doi.org/10.1109/TITB.2012.2227335

20. Chen X, Li Y (2013) A low power automatically warning system based on wireless sensor network. J Commun 8(8):512–520. https://doi.org/10.12720/jcm.8.8.512-520
21. Baccour N, Koubâa A, Mottola L, Zúñiga MA, Youssef H, Boano CA, Alves M (2012)Radio link quality estimation in wireless sensor networks: a survey. ACM Trans Sens Netw 8(4):1–33. https://doi.org/10.1145/2240116.2240123
22. Texas Instruments (2018) CC1310 SimpleLink™ Ultra-low-power Sub-1 GHz wireless MCU. Texas Instruments
23. Salgadoe S, Lu F, Lu F (2019) An anomaly detection model for ultra low powered wireless sensor networks utilizing attributes of IEEE 802.15.4e/TSCH. J Commun 14(5):335–341. https://doi.org/10.12720/jcm.14.5.33514.5.335-341

Satellite Communication Network and Signal Transmission

A Novel Multi-satellite Tracking Algorithm for the LEO Satellite Communications

Mengyue Liu, Haotian Chen, Zixian Ma, Bing Lan, Chunyi Song, and Zhiwei Xu

Abstract In the low earth orbit (LEO) satellite communication (SATCOM) scenario, the rough position of the satellite can be calculated from the ephemeris data. However, the multi-satellite tracking (MST) faces great challenges due to the inaccuracy of the ephemeris and interference of various sources. The traditional MST methods based on association suffer from high computational complexity, low estimation precision of satellite position and number. This paper then proposes an adaptive strong tracking unscented Kalman (ASTUK) MST algorithm based on probability hypothesis density (PHD), which reduces the computational complexity by setting the measurements partition strategy to optimally allocate the measurement data in the PHD filtering process, and improves the accuracy of state estimation by using the fading memory index (FMI) to adaptively update the information covariance of the gain matrix. Numerical simulations show the advantage of the proposed algorithm over the conventional ones on the computational complexity, tracking precision and estimation of satellite number.

Keywords Multi-satellite tracking · LEO satellite communication · Strong tracking unscented Kalman filter · Probability hypothesis density

M. Liu · H. Chen · Z. Ma · B. Lan · C. Song (✉) · Z. Xu
Institute of Marine Electronic and Intelligent System, Ocean College, Zhejiang University, Hangzhou, China
e-mail: cysong@zju.edu.cn

C. Song · Z. Xu
Engineering Research Center of Oceanic Sensing Technology and Equipment, Ministry of Education, Zhoushan, China

Donghai Laboratory, Zhoushan 316021, China

M. Ma (ed.), *Proceedings of the 12th International Conference on Communications, Circuits, and Systems*, Lecture Notes in Electrical Engineering 1193,
https://doi.org/10.1007/978-981-97-2636-3_4

1 Introduction

The low earth orbit (LEO) satellite communication (SATCOM) have recently gained broad research interests due to the less stringent requirements on power consumption and transmission signal delays [1]. The high-performance multi-satellite tracking (MST) for fast link switch is of great importance in guaranteeing the link quality of LEO SATCOM [2]. However, the influence of ephemeris data error and communication path clutter make it difficult to obtain accurate real-time data of the satellite. Single satellite tracking has made many considerable achievements relying on the development of phased array technology [3–5]. In order to improve the efficiency of MST, data fusion technology is applied to SATCOM in [6–8].

The commonly used multi-target tracking algorithm can realize MST by adapting to SATCOM scenario, which mainly includes the nearest neighbor data association (NNDA) [9], joint probabilistic data association (JPDA) [10], probabilistic multiple hypothesis tracking (PMHT) [11], and multiple hypothesis tracking (MHT) [12]. However, these algorithms are often used in two-dimensional target tracking, and there are some problems such as low accuracy and high computational complexity [13] when applied to MST. In our recent work, a MST algorithm based on PMHT and satellite orbit association (SOA) is proposed in [14]. This algorithm suffers from computational explosion and accumulated errors due to the use of data association [15]. To solve the above problems, the probability hypothesis density (PHD) algorithm is applied in algorithm, which uniformly model the measurement and the number and state of the target, which can effectively remove the clutter and avoid the correlation between the measured data and the tracked target, and estimate the number and state of the target directly by calculating the integral.

In this paper, an adaptive strong tracking unscented Kalman (ASTUK) algorithm based on PHD is proposed [16, 17]. The algorithm uses the probability hypothesis density (PHD) filter to transform the multi-target tracking problem into the set-valued estimation problem, and uses the measurements partition strategy to divide the measurement data into survival targets and new targets, which eliminate invalid measurements as much as possible and thus reduces the computational complexity, and improves the accuracy of multi-star tracking by designing and adopting adaptive fading memory index (FMI). The PHD filter extracts the multi-target state estimations from a posterior probability density and serves them as the prior value of the next moment. The numerical results verify the superior performance of the proposed algorithm.

The rest of this paper is organized as follows. Section 2 briefly reviews the system model and problem formulation. In Sect. 3, the proposed algorithm is introduced for MST. Experimental results including comparisons are presented in Sect. 4. Finally, the conclusion is given in Sect. 5.

2 System Model and Problem Formulation

2.1 The Satellite Tracking Model

The position of the satellite can be obtained from the direction of arrival (DOA) of the satellite signal estimated by the phased array. In our work, It is assumed that the DOA of the satellite can be observed by the phased array, and the Gaussian mixture PHD filter is used to track the satellite motion. The state x_k and the measurement z_k are given by

$$x_k = \varphi_k(x_{k-1}, \omega_{k-1}) \tag{1}$$

$$z_k = h_k(x_k, v_k) \tag{2}$$

where φ_k and h_k are the known nonlinear functions, ω_{k-1} and v_k are zero-mean Gaussian process noise and measurement noise with covariances Q_{k-1} and R_k, respectively. For satellites, $x_k = \left[y_k^T, \zeta_k^T\right]^T$, ζ_k^T is the turn rate and $y_k = \left[p_{x,k}, p_{y,k}, \dot{p}_{x,k}, \dot{p}_{y,k}, p_{z,k}, \dot{p}_{z,k}, \dot{\zeta}_k\right]^T$. In the satellite tracking scenario, due to the short sampling interval, the state and measurement formula can be approximated to

$$\mathrm{x_k} = F(\mathrm{\zeta_{k-1}})\mathrm{x_{k-1}} + \Gamma_\mathrm{k}\omega_\mathrm{k-1} \tag{3}$$

$$\zeta_k = \zeta_{k-1} + T\epsilon_{k-1} \tag{4}$$

$$z_k = H(x_k) + v_k \tag{5}$$

where

$$F(k) = \begin{bmatrix} 1 & a & 0 & -b & 0\,0\,0 \\ 0 & \cos\zeta T & 0 & -\sin\zeta T & 0\,0\,0 \\ 0 & b & 1 & a & 0\,0\,0 \\ 0 & \sin\zeta T & 0 & \cos\zeta T & 0\,0\,0 \\ 0 & 0 & 0 & 0 & 1\,T\,0 \\ 0 & 0 & 0 & 0 & 0\,0\,1 \\ 0 & 0 & 0 & 0 & 0\,0\,1 \end{bmatrix} \tag{6}$$

$$\Gamma_k = \begin{bmatrix} \frac{1}{2}aT^2 & 0 & 0 & 0 \\ T & 0 & 0 & 0 \\ 0 & \frac{1}{2}aT^2 & 0 & 0 \\ 0 & T & 0 & 0 \\ 0 & 0 & \frac{1}{2}aT^2 & 0 \\ 0 & 0 & T & 0 \\ 0 & 0 & 0 & \zeta T \end{bmatrix} \tag{7}$$

$$H(k) = \begin{bmatrix} 1\,0\,0\,0\,0\,0\,0 \\ 0\,0\,1\,0\,0\,0\,0 \\ 0\,0\,0\,0\,1\,0\,0 \end{bmatrix} \tag{8}$$

$$\begin{aligned} A &= \arctan\left(\frac{p_{y,k}}{p_{x,k}}\right) \\ E &= \arctan\left(\frac{p_{z,k}}{\sqrt{p_{x,k}^2+p_{y,k}^2}}\right) \end{aligned} \tag{9}$$

where T is the sampling interval, $a = \sin\zeta T/\zeta$, $b = (1 - \cos\zeta T)/\zeta$, ω_k and ϵ_k satisfy Gaussian distribution. A denotes the azimuth angle and E denotes the Pitch angle. The Fig. 1 illustrates the satellite motion in the local tangent coordinate system.

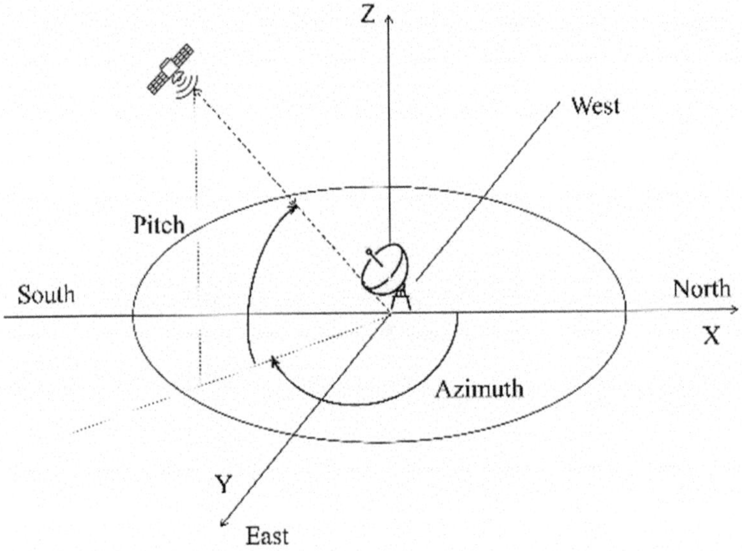

Fig. 1 Satellite motion in the local tangent coordinate system

2.2 PHD Estimation Model

In a multi-target scenario, targets appear and disappear randomly, the multi-satellite states and measurements can be characterized by random finite sets (RFS).

$$\begin{aligned} X_k &= \{x_{k,1}, \ldots, x_{k,n_k}\} \in F(\mathcal{X}) \\ Z_k &= \{x_{k,1}, \ldots, x_{k,n_k}\} \in F(\mathcal{Z}) \end{aligned} \tag{10}$$

where n_k is the number of targets, m_k is the number of measurements at time k. $F(\mathcal{X})$ and $F(\mathcal{Z})$ are the set of all possible finite subsets of state space $\mathcal{X}$ and measurement space $\mathcal{Z}$ respectively.

The function of PHD is a multimodal function in the single target state space. The number of peaks is the number of targets, and the location of peaks is the expected value of each target state. The recursive process of the PHD filter has two steps: prediction and update.

(1) Prediction: Suppose that the posterior PHD function at time $k-1$ is $D_{k-1}(x_{k-1})$, the predicted PHD function is given by

$$\begin{aligned} D_{k|k-1}(x_k) = \gamma(x_k) + \int & p_{s,k}(x_{k-1}) f_{k|k-1}(x_k|x_{k-1}) \\ & + \beta_{k|k-1}(x_k|x_{k-1}) D_{k-1} dx_{k-1} \end{aligned} \tag{11}$$

where $\gamma(x_k)$ is the intensity of spontaneous birth model, $p_{s,k}(x_{k-1})$ is the probability of target survival, $f_{k|k-1}(x_k|x_{k-1})$ is Markov transition density, and $\beta_{k|k-1}(x_k|x_{k-1})$ is the intensity of spawned target with state x_{k-1} at time k.

(2) Update: The predicted PHD $D_{k|k-1}(x_k)$ is updated with the measurements Z_k obtained at time k. Then, the updated PHD is given by

$$\begin{aligned} D_k(x_k) &= \left(1 - P_{D,k}(x_k) D_{k|k-1}(x_k)\right) \\ &+ \sum_{z \in Z_k} \frac{P_{D,k}(x_k)\varphi_k(z|x_k) D_{k|k-1}(x_k)}{\lambda_c c(z) + \int P_{D,k}(x_k)\varphi_k(z|\xi) D_{k|k-1}(\xi) d\xi} \end{aligned} \tag{12}$$

where $P_{D,k}(x_k)$ is the detection probability, λ_c is the Poisson distributed with the average rate of the clutter numbers, $c(z)$ is the probability density of the spatial distribution of clutters, and $\varphi_k(z|\xi)$ is the measured likelihood function.

3 The Proposed Algorithm

3.1 Improved MST Algorithm Based on PHD

In satellite tracking scene, measurement is easily affected by clutter, which leads to inaccurate mapping between measurement data and tracking data. This algorithm unifies the randomness of measurement and target number and state, avoiding complex data association problems.

Step 1: Initial Prior

Given $\left\{W_{k-1}^{(i)}, m_{k-1}^{(i)}, P_{k-1}^{(i)}\right\}_{i=1}^{J_{k-1}}$, and the measurement setsZ_k, where $W_{k-1}^{(i)}, m_{k-1}^{(i)}, P_{k-1}^{(i)}$ are the weight, mean and covariance of the posterior probability density respectively, J_{k-1} is the number of Gaussian term at time$k$. Initial satellite measurements are generally calculated from ephemeris data.

Step2: Prediction Process

$$\mu = \begin{bmatrix} m_{k|k-1}^{(i)} \\ 0 \\ 0 \end{bmatrix} \tau = \begin{bmatrix} P_{k|k-1}^{(i)} & 0 & 0 \\ o & Q_{k|k-1}^{(i)} & 0 \\ 0 & 0 & R_k^{(i)} \end{bmatrix}$$

Use the unscented transformation with mean μ and covariance τ to generate L sigma points and weights. $l = 0, \ldots, L$ and $i = 0, \ldots, J_{k-1}$ is the number of sigma points and Gaussian terms, respectively.

$$x_{k|k-1}^{(i)} = \sum_{l=0}^{L} W^{(l)} \varphi\left(x_{k-1}^{(l)}, w_{k-1}^{(l)}\right) \tag{13}$$

$$m_{k|k-1}^{(i)} = \sum_{l=0}^{L} W^{(l)} x_{k|k-1}^{(l)} \tag{14}$$

$$P_{k|k-1}^{(i)} = \sum_{l=0}^{L} W^{(l)} \left(x_{k|k-1}^{(l)} - m_{k|k-1}^{(i)}\right)\left(x_{k|k-1}^{(l)} - m_{k|k-1}^{(i)}\right)^T \tag{15}$$

Step 3: Update Progress

The measurements are divided into survival target measurements Z_k^S and newborn measurements Z_k^γ including new target measurements and clutter measurements by using the measurements partition strategy. Then Z_k^S is used in update progress. The lth Gaussian measurement and ith sigma point measurement are given by

$$z_{k|k-1}^{(s,l)} = h\left(x_{k-1}^{(l)}, W_{k-1}^{(l)}\right) \tag{16}$$

$$\eta_{k|k-1}^{(s,l)} = \sum_{l=0}^{L} W^{(l)} z_{k|k-1}^{(s,l)} \tag{17}$$

In order to reasonably balance the contribution of prediction information and observation information to filtering estimation, and reduce the impact of abnormal state parameter estimation and model abnormal error on filtering estimation at the previous time, the FMI weighting α_k is introduced in the update progress

$$P_{k|k}^{(i)} = \alpha_k P_{k|k}^{(i)} - G_k^{(i)} \left[S_k^{(i)}\right]^{-1} \left[G_k^{(i)}\right]^T \tag{18}$$

$$W_{k|k}^{(i)} = P_{D,k} w_{k|k-1}^{(i)} N\left(x_{k|k-1}^{(l)}; \eta_{k|k-1}^{(i)}, S_{k|k-1}^{(i)}\right) \tag{19}$$

$$m_{k|k}^{(i)} = m_{k|k-1}^{(i)} + K_k^{(i)}\left(z_{k|k-1}^{(s,l)} - \eta_{k|k-1}^{(i)}\right) \tag{20}$$

where

$$S_k^{(i)} = \sum_{l=0}^{L} W^{(l)}\left(z_{k|k-1}^{(s,l)} - \eta_{k|k-1}^{(i)}\right)\left(z_{k|k-1}^{(s,l)} - \eta_{k|k-1}^{(i)}\right)^T + R_k \tag{21}$$

$$G_k^{(i)} = \sum_{l=0}^{L} W^{(l)}\left(x_{k|k-1}^{(s,l)} - m_{k|k-1}^{(i)}\right)\left(z_{k|k-1}^{(s,l)} - \eta_{k|k-1}^{(i)}\right)^T \tag{22}$$

$$K_k^{(i)} = G_k^{(i)}\left[S_k^{(i)}\right]^{-1} \tag{23}$$

where $P_{k|k-1}^{(i)}$ and $P_k^{(i)}$ denote the predicted covariance and the updated covariance of the states at time k, $G_k^{(i)}$ denotes the cross covariance of state and measurement, $S_k^{(i)}$ denotes the covarance of measurement and $K_k^{(i)}$ denotes the Kalman gain.
Step 4: State estimation
By tracking the target components corresponding to the weights above a specific threshold and ignoring other components, a posteriori intensity is used to calculate the tracking multi-target states $\widehat{X}_k = \left\{m_k^{(i)} : W_k^{(i)} > 0.5\right\}$ and the numbers $\widehat{N}_k$.

3.2 *Measurement Partition Strategy*

In order to reduce the computational complexity, a window strategy is constructed to filter out and establish the new target measurement set Z_k^{γ} and the survival target measurement set Z_k^S, and only the survival target measurement set is used to participate in the update process. Let

$$\Phi_k(z) = \left\{\phi_k^{(1)}(z), \ldots, \phi_k^{(l)}(z), \ldots, \phi_k^{(\gamma_k)}(z) | z \in Z_k\right\} \tag{24}$$

where $\phi_k^{(i)}(z) = (\Delta z_k)^T S_k^{(i)} \Delta z_k$, γ_k is the number of Gaussian term. Let

$$\Delta z_k = z_{k|k-1}^{(l)} - \eta_{k|k-1}^{(i)} \tag{25}$$

be the residual error of the lth Gaussian term and the ith measurement. Then the covariance matrix corresponding to the residual vector can be expressed as $S_k^{(i)}$.

The measurements partition strategy is given by

$$(\Delta z_k)^T S_k^{(i)} \Delta z_k \leq T_k^{(1)} \tag{26}$$

$$T_k^{(l)} = 2ln\left[\frac{P_g W^{(l)}}{(1-p_g)\kappa_k\sqrt{\left|S_k^{(i)}\right|}}\right] \tag{27}$$

$$Z_k = \begin{cases} Z_k^{\gamma} = \left\{z_{\gamma} \in Z_k | (\Delta z_k)^T S_k^{(i)} \Delta z_k > T_k^{(l)}\right\} \\ Z_k^{s} = \left\{z_s \in Z_k | (\Delta z_k)^T S_k^{(i)} \Delta z_k \leq T_k^{(l)}\right\} \end{cases} \tag{28}$$

where $T_k^{(l)}$ is the threshold of the $l - th$ Gaussian component, P_g is the probability of correctly falling within the $T_k^{(l)}$, κ_k is the clutter density.

3.3 FMI Weighting

When the model and process parameters are suitable, the traditional tracking algorithm converges gradually and the unbiased estimation of the state can be obtained. However, when the model mismatch is large, the estimation accuracy will greatly decrease or even diverge. In this subsection, we introduce the FMI weighting to adjust the predicted state covariance in real time. The actual residual variance of the system.

$$\Lambda z_k = z_k - \eta_{k|k-1}^{(i)} \tag{29}$$

Under the adjustment of FMI weighting α_k, it should be satisfied $\Lambda z_k = \Delta z_k$, the α_k can be calculated as follows:

$$\alpha_k = \begin{cases} \alpha_0 & \alpha_0 \geq 1 \\ 1 & \alpha < 1 \end{cases} \tag{30}$$

$$\alpha_0 = \frac{tr\,\Xi_k}{tr\mathscr{B}_k} \tag{31}$$

Based on the above method, the following formulas give the equivalent representation of Ξ_k and $\mathscr{B}_k$.

$$\Xi_k = C_{0,k} - \psi R_k \tag{32}$$

$$\mathscr{B}_k = \sum_{l=0}^{L}\left(z_{k|k-1}^{(s,l)} - \eta_{k|k-1}^{(i)}\right)\left(z_{k|k-1}^{(s,l)} - \eta_{k|k-1}^{(i)}\right)^T \tag{33}$$

$$C_{0,k} = \begin{cases} \Delta z_k(\Delta z_k)^T k = 1 \\ \frac{\rho C_{0,k-1}\Delta z_k(\Delta z_k)^T}{1+\rho} k \geq 2 \end{cases} \tag{34}$$

where $0 < \rho \leq 1$ denotes the fading factor, $\psi \geq 1$ denotes the determined the attenuator factor.

3.4 Algorithm Description

Table 1 is the flow of the proposed ASTUK algorithm.

4 The Proposed Algorithm

In this example, We use Matlab R2021b on a PC with a 2.90 GHz Intel Core i5 and 16 GB of RAM to conduct all experiments. Each target has a survival probability $P_{S,k} = 0.99$ and dectection probability $P_{D,k} = 0.98$. The formulas of the model are given in Sect. 2-A *The Satellite Tracking Model*. $T = 0.1s$, $\omega_k \sim N(\cdot; 0, \sigma_\omega^2 I_2)$, $\sigma_\omega = 15\,\text{m/s}^2$, and $\epsilon_k \sim N(\cdot; 0, \sigma_\epsilon^2)$, $\sigma_\mu = \left(\frac{\pi}{180}\right)$ rad/s, $J_{\max} = 100$. Deviation of process noise $\sigma_Q^2 = 1\,\text{rad}^2$, deviation of measurement noise $\sigma_R^2 = 1\,\text{rad}^2$. The constellation is configured as a walker constellation with the number of satellites $N = 720$, the orbit planes $P = 12$, and the harmonic factor $m = 0$. The estimated number of interested targets and the optimal sub-pattern assignment (OSPA) distance are used to evaluate the tracking performance by the Monte-Carlo simulation with simulation time $I = 500$. A total of 10 satellites were simulated, the change in the number of which can be seen in Figs. 3 and 6. The start time and end time were given randomly to simulate the uncertainty of multiple satellites tracked by different ground stations. The clutter RFS follows the uniform Poisson model over the surveillance region of $[A, E]$ is $[-\pi, \pi; \pi/6, \pi/2]$, with $\lambda_c = 15\,(\text{radm})^{-1}$, an average of 20 clutters in the surveillance region.

Table 1 The proposed Astuk algorithm

Algorithm	Description of the proposed MST algorithm
1:	Initiate tracking state paramaters from Empheris data
2:	Given $\left\{\boldsymbol{W}_{k-1}^{(i)}, \boldsymbol{m}_{k-1}^{(i)}, \boldsymbol{P}_{k-1}^{(i)}\right\}_{i=1}^{J_{k-1}}$, and the measurement sets $\boldsymbol{Z}_k$
3:	**repeat**
4:	Generate the sigma points via μ and τ
5:	Prediction to calculate $\boldsymbol{x}_{k-1}^{(i)}, \boldsymbol{m}_{k-1}^{(i)}, \boldsymbol{P}_{k-1}^{(i)}$
6:	Construction of PHD update components to calculate $\boldsymbol{z}_{k\|k-1}^{(i)}$, $\boldsymbol{\eta}_{k\|k-1}^{(i)}, \boldsymbol{P}_{k\|k}^{(i)}, \boldsymbol{W}_{k\|k}^{(i)}, \boldsymbol{m}_{k\|k}^{(i)}, \boldsymbol{S}_k^{(i)}, \boldsymbol{G}_k^{(i)}, \boldsymbol{K}_k^{(i)}$
7:	Establish the newborn measurement set $\boldsymbol{Z}_k^{\gamma}$ and the survival target measurement set $\boldsymbol{Z}_k^{s}$ by measurements partition strategy
8:	The $\boldsymbol{Z}_k^{s}$ is used in update progress
9:	Introduce FMI weighting and calculate the residual error $\boldsymbol{\Delta z}_k$ and parameters of FMI: $\boldsymbol{\alpha}_k$, $\boldsymbol{\alpha}_0$, $\boldsymbol{\Xi}_k$, $\boldsymbol{\mathcal{B}}_k$, $\boldsymbol{C}_{0,k}$
10:	Calculate $\boldsymbol{S}_k^{(i)}, \boldsymbol{G}_k^{(i)}, \boldsymbol{P}_{k\|k}^{(i)}$ under FMI wweighting to replace the the old ones
11:	Update progress to calculate $\boldsymbol{P}_{k\|k}^{(i)}, \boldsymbol{W}_{k\|k}^{(i)}, \boldsymbol{m}_{k\|k}^{(i)}$
12:	Calculate the tracking target states $\widehat{\boldsymbol{X}}_k$ and the numbers $\widehat{\boldsymbol{N}}_k$
13:	Substitute the output of 11 into 2 to start the new calculation
14:	**Until** Order to stop the tracking

This subsection is divided into two experiments: multi-satellite tracking without interference and multi-satellite tracking under interference. The compared algorithms are the SOA MST algorithm proposed in [14], the PMHT algorithm [11] the Extension Kalman PHD filter (EKF-PHD) [18] and the Sequential Monte Carlo PHD filter (SMC-PHD) [19]. The average runtime for one sampling time of the proposed algorithm and others in experiment A and B are given in Fig. 8, which shows that the proposed algorithm has the best MST speed in both experiment A and B, followed by SOA algorithm. The SMC-PHD uses a large number of particle samples to numerically approximate the posterior probability density, and the computational complexity and runtime are relatively high. The PMHT algorithm achieves MST by optimizing the track-echo correlation and smooth estimation through a finite number of iterative loops. With the increase of the number of satellites, the amount of computation increases linearly with the number of targets and measurements, so its runtime is the highest of all MST algorithms.

4.1 Multi-satellite Tracking Without Interference

The Fig. 2 shows the tracking trajectories of the proposed algorithm in experiment A. In Fig. 3, the performance comparisons of the proposed algorithm and the other MST algorithms are carried out in terms of the mean estimated target number when tracking satellites without interference of other uninterested targets. We can conclude that the proposed algorithm has excellent performance. In Fig. 4, the proposed algorithm also has a superior performance on OSPA distance [20], followed by the SOA. The OSPA of EKF-PHD increase gradually due to the filter divergence caused by the inaccurate estimation of noise covariance matrix, the OSPA of the SMC-PHD and PMHT fluctuates greatly and cannot be maintained at an accurate accuracy level.

It is found that the proposed algorithm has a more significant performance when target numbers change over time, and has an effective and stable target tracking performance campared with other MST algorithms.

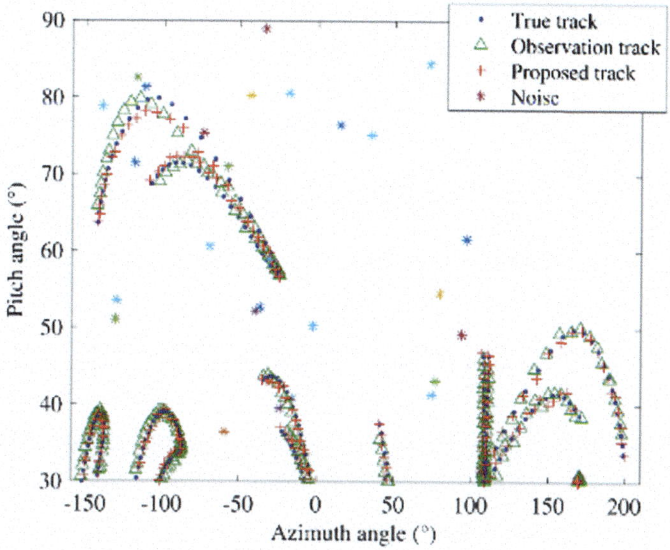

Fig. 2 Tracking trajectories of the proposed algorithm without interference

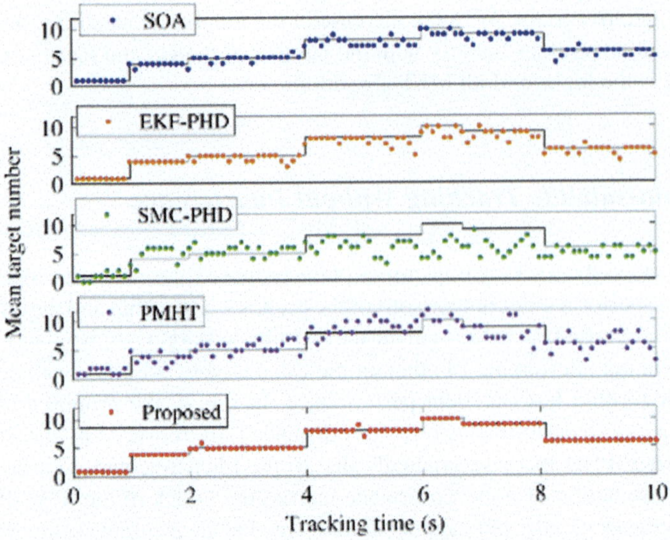

Fig. 3 Tracking performance on mean target number without interference

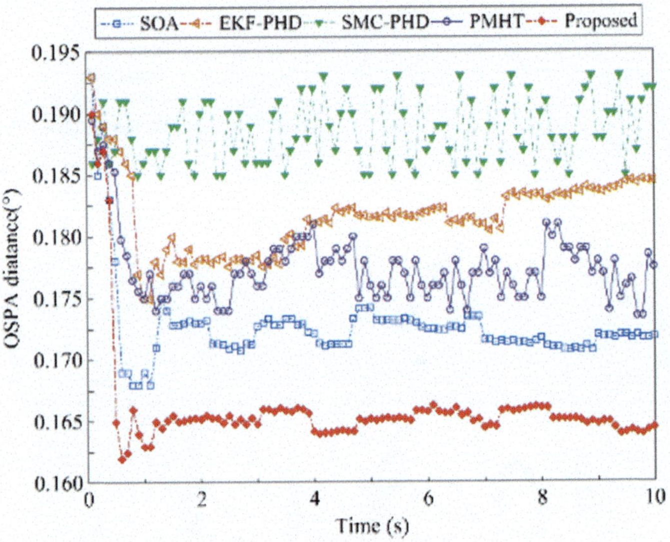

Fig. 4 Tracking performance on OSPA distance without interference

4.2 Multi-satellite Tracking Under Interference

There are two interference tracks in experiment B and the tracking trajectories of the proposed algorithm is shown in Fig. 5. In Figs. 6 and 7, the performance comparisons of the proposed algorithm and the other MST algorithms are carried out in terms of the mean estimated target number and the OSPA distance. We can see form these figures that at t = 4 s, the tracking performance of all algorithms in these two aspects degrades to some extent due to the influence of the two interference tracks. In particular, the performance of the PMHT algorithm deteriorates seriously because it adopts a measurement-oriented parameter model, which is easily disturbed by clutter. However, the proposed algorithm can recover quickly and reach the accuracy level before interference and which shows that it has good anti-interference ability and fast recovery ability. In addition, compared with other algorithms, the proposed algorithm has less error fluctuation and better stability.

The tracking performance of the PMHT, EKF-PHD and SMC-PHD gradually deteriorates to some extent with the increase of time. Among those, the EKF-PHD suffers the most due to the filtering divergence, which not only leads to the gradual increase of OSPA value, but also produces a large error in the estimation of the number of tracking targets. The SOA has better anti-interference ability, but it is not excellent in tracking accuracy. However, the tracking performance of the proposed algorithm has always remained stable and superior, which proves that the proposed algorithm has better MST performance than other algorithms.

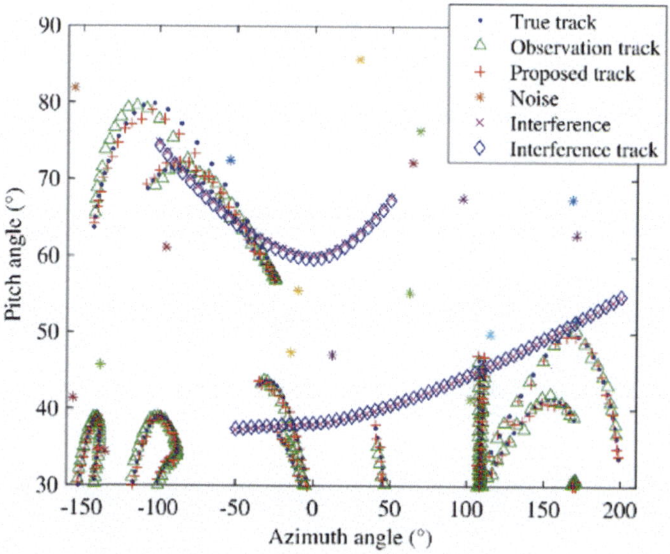

Fig. 5 Tracking trajectories of the proposed algorithm under interference

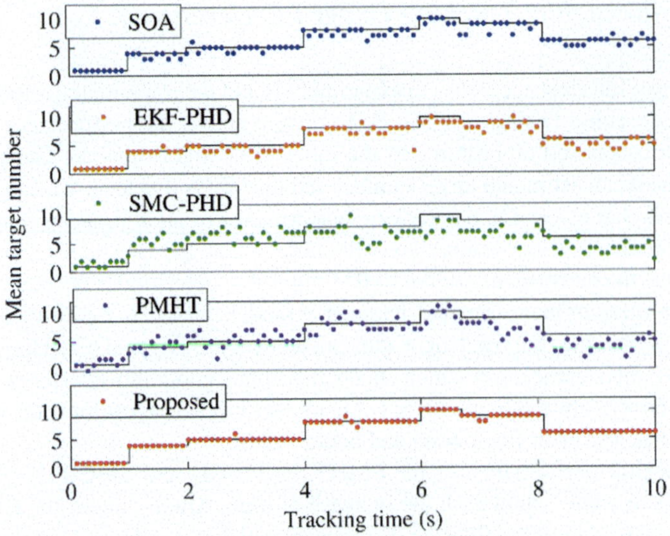

Fig. 6 Tracking performance on mean target number under interference

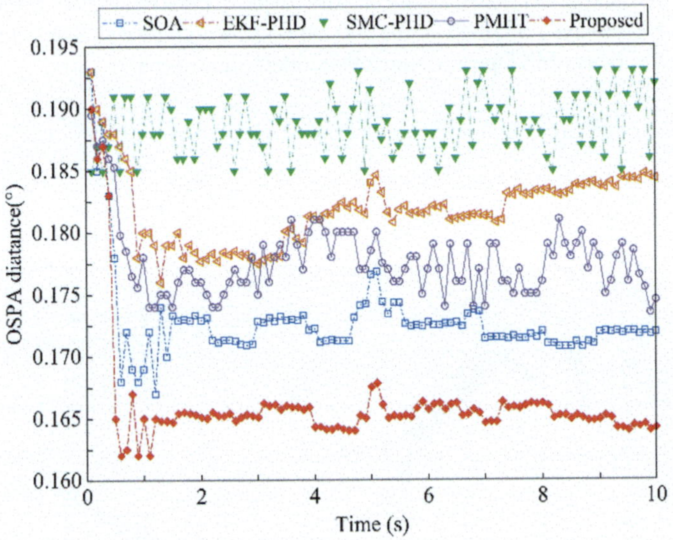

Fig. 7 Tracking performance on OSPA distance under interference

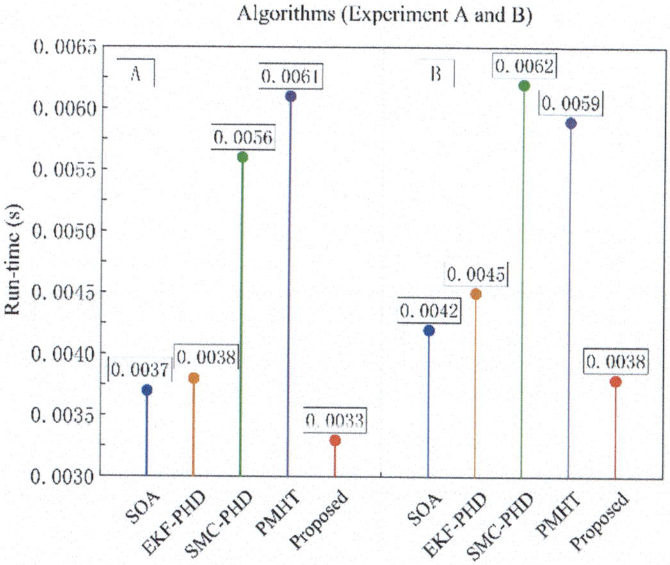

Fig. 8 Runtime of algorithms in experiment A and B

5 Conclusion

The fast and accurate multi-satellite tracking (MST) is of great significance to LEO SATCOM. An adaptive strong tracking unscented Kalman (ASTUK) algorithm based on probability hypothesis density (PHD) is proposed for the MST in this paper. The algorithm reduces the computational complexity by establishing a measurements partition strategy to optimally allocate the measurement data, and introduces fading memory index (FMI) to adaptively adjust the covariance matrix to improve tracking accuracy. The numerical simulation results verify that the proposed algorithm simultaneously obtains decreased computational complexity and improved tracking accuracy, as compared to the conventional algorithms.

References

1. Sağ E, Kavas A (2018) Modelling and performance analysis of 2.5 Gbps inter-satellite optical wireless communication (IsOWC) system in LEO constellation. J Commun 13(10):553–558
2. Bayer H, Volmer C, Krauß A, Stephan R, Hein MA (2010)Tracking antenna for mobile bi-directional satellite communications in Ka-band. In: 2010 IEEE international conference on wireless information technology and systems
3. Bai X, Liao C, Pan X, Xu M (2019) Mining two-line element data to detect orbital maneuver for satellite. IEEE Access 7:129537–129550

4. Son SH, Eom SY, Jeon SI (2003) A novel tracking control realization of phased array antenna for mobile satellite communications. In: The 57th IEEE semiannual vehicular technology conference, vol 4, pp 2305–2308
5. (2019) Adaptive tracking for beam alignment between ship-borne digital phased-array antenna and LEO satellite. J Commun Inf Netw 4(3):60–70
6. Hongying C, Caifa G, Xuliang W, Huifen L (2015)Research of multi-satellite tracking algorithm based on data fusion. In: 2015 IEEE international conference on signal processing, communications and computing (ICSPCC), pp 1–4
7. Wang H, Yi J, Wan X (2018) Tracking algorithm with data fusion in single frequency network-based MISO passive radar. In: 2018 12th international symposium on antennas, propagation and EM theory (ISAPE), pp 1–3
8. Zhao S, Huang Y, Wang K, Chen T (2021) Multi-source data fusion method based on nearest neighbor plot and track data association. IEEE Sens 2021:1–4
9. Singer RA, Stein JJ (1971) An optimal tracking filter for processing sensor data of imprecisely determined origin in surveillance systems. In: 1971 IEEE conference on decision and control, pp 171–175
10. Fortmann T, Bar-Shalom Y, Scheffe M (1983) Sonar tracking of multiple targets using joint probabilistic data association. IEEE J Oceanic Eng 8(3):173–184
11. Willett P, Ruan Y, Streit R (2002) PMHT: problems and some solutions. IEEE Trans Aerosp Electron Syst 38(3):738–754
12. Blackman SS (2016) Abstracts of previous tutorials in this series: multiple hypothesis tracking for multiple target tracking. IEEE Aerosp Electron Syst Mag 31(3):90–96
13. Oussalah M, Schutter JD (2002) Hybrid fuzzy probabilistic data association filter and joint probabilistic data association filter. Inf Sci 142(1–4):195–226
14. Chen Q, Ma Z, Lan B, Song C, Xu Z (2022)Multi-satellite tracking for the LEO satellite communication network. In: ICC 2022—IEEE international conference on communications, Seoul, Korea, Republic of China, pp 3082–3087
15. Houssineau J, Laneuville D (2010)PHD filter with diffuse spatial prior on the birth process with applications to GM-PHD filter. In: 2010 13th international conference on information fusion, pp 1–8
16. Gourabi FN, Kiani M, Pourtakdoust SH (2021) On-line orbit and albedo estimation using a strong tracking algorithm via satellite surface temperature data. IEEE Trans Aerosp Electron Syst 57(3):1443–1454
17. Mahler RPS (2003) Multitarget Bayes filtering via first-order multitarget moments. IEEE Trans Aerosp Electron Syst 39(4):1152–1178
18. Wang X, Wang J (2016)Simulation analysis of EKF and UKF implementations in PHD filter. In: 2016 IEEE 13th international conference on networking, sensing, and control (ICNSC), pp 1–6
19. Liu Z, Wang Z, Xu M, Yang L, Liu J (2015)A novel CIF-based SMC-PHD approach for tracking multiple nonlinear targets. In: 2015 IEEE radar conference (RadarCon), pp 0555–0560
20. Beard M, Vo BT, Vo B-N (2017) OSPA$^{(2)}$: using the OSPA metric to evaluate multi-target tracking performance. In: 2017 international conference on control, automation and information sciences (ICCAIS), pp 86–91

A Link Switching and Traffic Scheduling Method for Satellite-Ground Uninterrupted Communication

Yiqiu Zhou, Kanglian Zhao, Wenfeng Li, and Yuan Fang

Abstract Space laser communications gradually reveal an increasing development trend due to the outstanding features of laser links. However, the satellite-ground laser link can be easily affected by the absorption, scattering, and turbulence of the atmospheric clouds. It will lead to the laser transmission signal being seriously attenuated or even interrupted, which becomes the bottleneck of the satellite-ground laser networking. In this paper, considering the ground station (GST) multiple access candidates and meteorological factors, a predictable network mechanism of satellite-ground laser links construction is proposed to ensure the continuous communications of the satellite-ground laser network. By introducing the prediction of the atmospheric change trend of a specified area in a certain period, we evaluate the performance of this networking approach in terms of probability of interruption, end-to-end delay, and other Quality-of-Service (QoS) indicators. The simulation results indicate that if we could ensure the accuracy of the prediction of meteorological factors over the GSTs, the corresponding GSTs will be selected for communication through the results. Therefore, the end-to-end networking system will achieve long-term uninterrupted communication in the future.

Keyword Satellite-ground laser networking · Predictable networking · Link switching · Traffic scheduling

---

Y. Zhou · K. Zhao (✉) · W. Li · Y. Fang
School of Electronic Science and Engineering, Nanjing University, Nanjing, China
e-mail: zhaokanglian@nju.edu.cn

Y. Zhou
e-mail: yqzhou@nju.edu.cn

W. Li
e-mail: leewf_cn@hotmail.com

Y. Fang
e-mail: yfang@nju.edu.cn

M. Ma (ed.), *Proceedings of the 12th International Conference on Communications, Circuits, and Systems*, Lecture Notes in Electrical Engineering 1193,
https://doi.org/10.1007/978-981-97-2636-3_5

1 Introduction

With the development of information society and the increase of spatial perception data, traditional microwave communication technology has been difficult to meet the needs of space networks for larger data transmission volume and faster transmission rate. Among the many solutions, laser communication can significantly increase the data rate of satellite networks, provide greater communication capacity, and relieve traffic pressure in satellite networks. Compared with traditional Radio Frequency (RF) communication, laser communication has the advantages of larger bandwidth, lower power, smaller terminal size, higher directivity and higher security. Therefore laser communication has become the first choice to meet the large-capacity and high-rate transmission requirements of future space networks.

However, in satellite-ground laser communication, laser transmission is greatly affected by atmospheric conditions, such as atmospheric turbulence, cloud rain and other factors, which will induce channel fading, satellite receiving information bit error or communication interruption, resulting in a decrease in the reliability of laser links. The challenges which laser link might encounter are presented in Fig. 1.

Focusing on solving the problem of instability of satellite-ground laser communication, relevant technologies can be used to reduce the influence of atmospheric factors. Current commonly used technical means include:

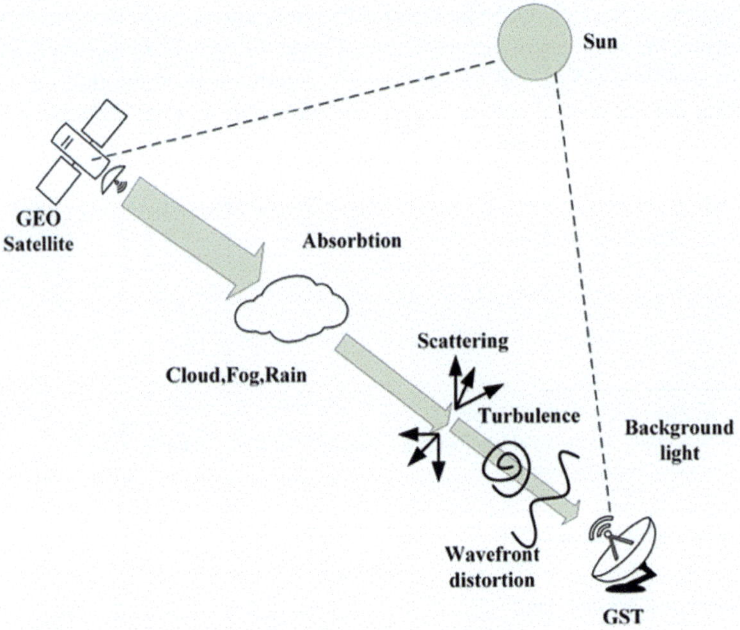

Fig. 1 Challenges for satellite-to-ground laser links

- **Site diversity**: It has been widely used to improve the availability of satellite-ground communication networks by establishing transmission links with multiple GSTs to reduce the adverse effects of cloud layers on links.
- **Adaptive optics**: the optical control method adopted in the physical layer pre-corrects the conjugate signal of atmospheric turbulence before atmospheric transmission so that the laser is transmitted through the atmosphere without distortion [1].
- **Network re-routing**: A routing mechanism adopted in the network layer that rediscovers routes for forwarding when a link goes down.

One commonly used physical layer technique is site diversity. Site diversity technique reduces the adverse effects of clouds on satellite-ground laser links by establishing transmission links with multiple optical GSTs, and has been widely used to improve the availability of satellite-ground LASER communication networks, that is, the probability that satellites and ground can establish communication. The traditional site diversity technology adopts the idea that a system will switch the link after it is not working [2]. In concrete terms, if the satellite-ground laser link is interrupted by atmospheric factors, it will induce the system to reroute to the available GST. After the rerouting process is completed, the system will resume work. During the route re-establishment period, the satellite-ground laser link will be interrupted, and in the scenario of satellite-ground backbones, if the link is interrupted, it will affect the transmission of high-speed, real-time services, causing immeasurable losses. Therefore, the mechanism called 'passive response' unique to the traditional site diversity technique is difficult to avoid the interruption of high-speed backbone links between satellites and grounds.

At present, there are many types of meteorological prediction technology methods, such as cloud change prediction, which belongs to the research scope of time series forecasting, mainly divided into statistical analysis method, numerical prediction method, optical flow prediction method and machine learning prediction method. The Hong Kong University of Science and Technology has conducted research on cloud trajectories [3]. After knowing the movement and changes of the cloud in advance, the GST can centrally allocate link resources and actively switch to the appropriate GST when there is a cloud blocking the link to avoid the link interruption problem caused by passive switching.

Based on the above forecasting techniques, we can quickly predict the meteorological conditions over the GST and evaluate communication quality indicators [4]. The research is based on the premise of short-term predictability of meteorological factors. Before the laser link is about to be interrupted, a GST with good link-building conditions is selected in advance to realize the switching and scheduling of reliable links and avoid the occurrence of a temporary outage. This is of great significance for the realization of long-term uninterrupted communication between satellites and grounds.

2 System Model

As mentioned above, our goal is to realize long-term uninterrupted communication between satellites and ground, and get rid of the constraints and influences of satellite-ground laser links by atmospheric factors. Although the technique of site diversity can improve the availability of GST network, the passive response mechanism in the traditional optical GST site diversity scheme will still cause the link to be interrupted, affecting high-speed services and incurring re-routing costs. In this chapter we present a link switching method based on short-term prediction to avoid temporary interruption of the link, guaranteeing for the uninterrupted transmission of satellite-ground laser communication. The present mechanism judges the link construction conditions over the GST by short-term prediction. It makes the link construction decision every time slot, combined with the basic idea of GST site diversity, to avoid unexpected communication interruptions.

2.1 *Predictive Network Architecture*

Figure 2 shows the overall architecture diagram of predictive networking. The backbone satellite 1 ~ 3 form a geosynchronous orbit satellite network that can be used to transmit data directly to the GST or as a relay transmission. GST 1 ~ 3 forms an optical GST site-diversity network. Key user 1 ~ 3 briefly summarize the working mechanism of the scene. When the system predicts that there are rain clouds blocking the laser link over the GST 1, the system dispatches the data traffic to key users 1 through topology optimization. When the system predicts that there are thin clouds over key user 2 that cause link attenuation, it will optimize the link through adaptive coding modulation. When the system predicts that there will be a thick cloud blocking the link over key user 3, it will let the backbone satellite 3 establish a link with key user 3 through link switching, and the data on satellite 2 will be transferred through traffic scheduling dispatched to satellite 3 and transmitted downlink to key user 3. The above reflects the ideas of topology optimization, traffic scheduling, link optimization and link switching, which constitutes a preliminary air-space-ground integrated prediction network.

2.2 *Basic Working Mechanism*

Considering a basic scenario for a simplified satellite-ground network, as shown in Fig. 3. The GSTs are connected to the ground data center. Under the condition that there is only one ground laser terminal on the satellite, a satellite can only connect to one GST in the corresponding coverage area for data transmission at a time [5].

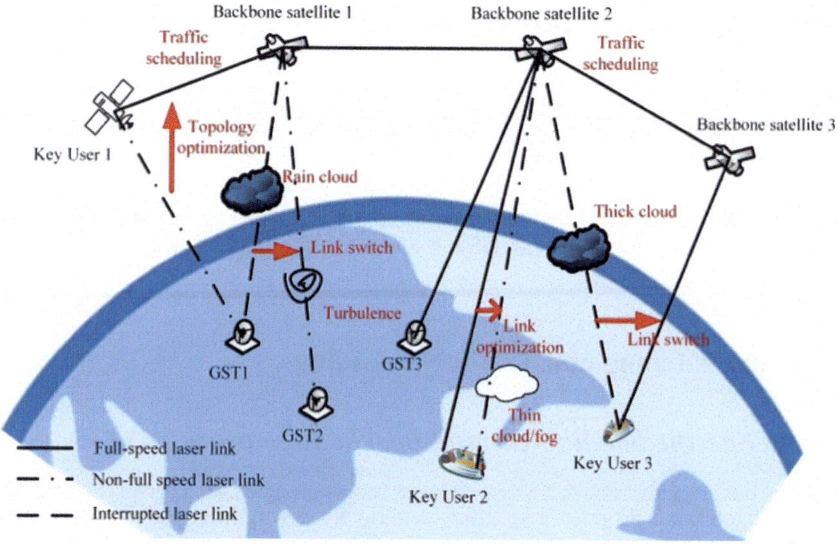

Fig. 2 The satellite-ground predictable networking architecture

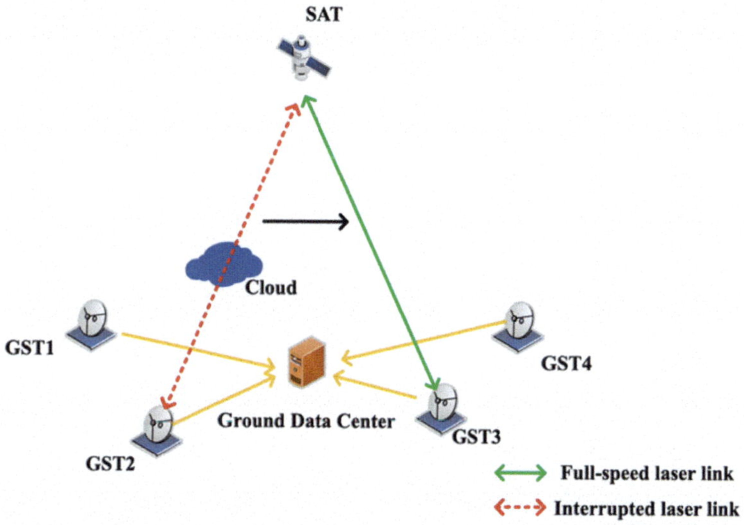

Fig. 3 A basic satellite-ground link switching system

Firstly, we introduce the influence of meteorological conditions on the laser link. If there exist clouds over the GST at a certain time, it can be regarded as the GST being completely covered by clouds so that any satellite cannot communicate with it [6].Define the state of communication between the GST and the satellite as a binary variable with a value of 1 if the link is obscured by clouds, otherwise a value of 0.

In addition, if the system have the short-term meteorological prediction results of all GSTs in the area, once the current link is going to be obscured by the clouds, the system will choose another good-condition GST to switch the link, the steps are as follows: the original satellite-ground link cut off the link, and then the satellite will rebuild the link with the new GST so that the data transmission is restored. As shown in Fig. 3, it is predicted that the sky above GST 2 will be obscured by clouds in the near future, while the weather conditions above GST 3 are good. At this time, satellite a will cut off the connection with GST 1, and then rebuild the link with GST 3.

3 Link Switching and Traffic Scheduling

3.1 A Specific Satellite-Ground Backbone Network

In this chapter, we propose a specific scenario for a simplified satellite-terrestrial backbone network, as shown in Fig. 4. The GEO satellites in the satellite-ground backbone network are in synchronous orbit (i.e. SAT a, SAT b, SAT c). Under the condition of stable inter-satellite link construction, the GEO backbone satellites can communicate with all the GSTs at any time. Therefore, GEO satellites are often used as relays for medium and low-orbit satellite constellations to achieve data sharing and realize the transmission of onboard data to GSTs in non-overhead scenarios.

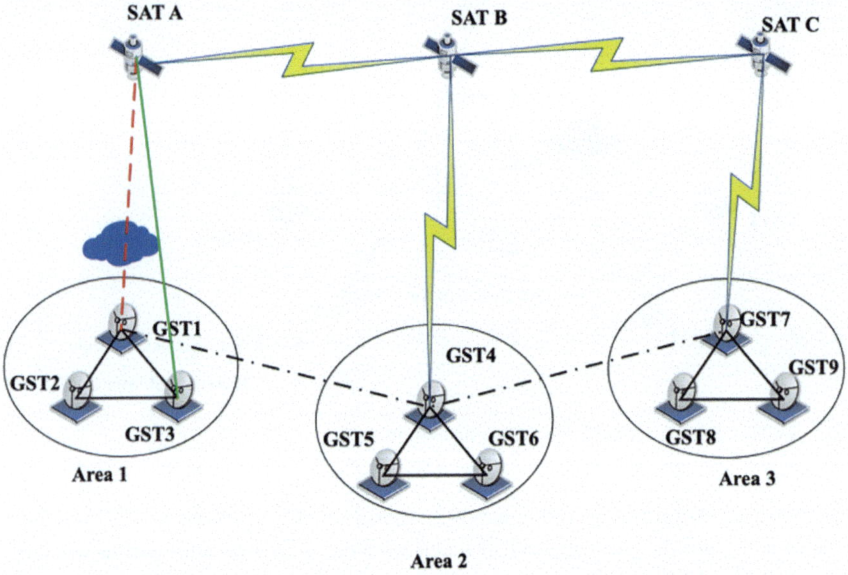

Fig. 4 A specific satellite-ground backbone network

GSTs connecting to the GEO satellites are deployed inside hosting network in separate areas. Each GEO satellite provides service to a certain area. For example, GEO A offers service to Area 1 which have deployed 3 GSTs (GST 1 ~ 3).

Under the condition that there is only one ground laser terminal on the GEO satellite, a GEO can only connect to one GST in the corresponding coverage area for data transmission at a time.

3.2 Link Switching

As shown in Fig. 5 is the time-slot setup in this scenario.

Given that the satellite-ground connection state of each time slot is determined and unique. Between the current time slot t_n and the next time slot t_{n+1}, a switching time slot $\Delta\tau$ is set for decision-making. In this time slot, any satellite-ground link can perform actions: Switch the GST to acquire, track and point (ATP) to build another link or maintain the current connection [7]. In the current link holding time slot $\Delta\varphi$, the system predicts the meteorological conditions over each GST in the next time slot, and evaluates the link construction conditions, so as to maintain the status quo or rebuild the link. A complete slot is defined as shown in Eq. (1).

$$\Delta t = \Delta\tau + \Delta\varphi = t_{n+1} - t_n \quad (1)$$

In each switching time slot, if the system chooses to switch the link, the steps are as follows.

If there exist available GST candidate in area 1 which means the meteorological parameters above it are within a certain interval, then the original satellite-ground link will be disconnected, finally the backbone satellite will rebuild the link with the new GST so that the data transmission is restored. As shown in Fig. 5, within time slot 1, it is predicted that the sky above GST 1 in area 1 will be obscured by clouds in the next time slot, while the weather conditions above GST 3 are good. At this time, satellite a will cut off the connection with GST 1, and then rebuild the link with GST 3. During this time slot, the network capacity of the system will have a certain loss, the loss capacity size is $\Delta\tau \cdot C_t$, where C_t, represents the communication capacity of the link within the time t [8]. At the beginning of time slot 2, the backbone link of area 1 resumes data transmission.

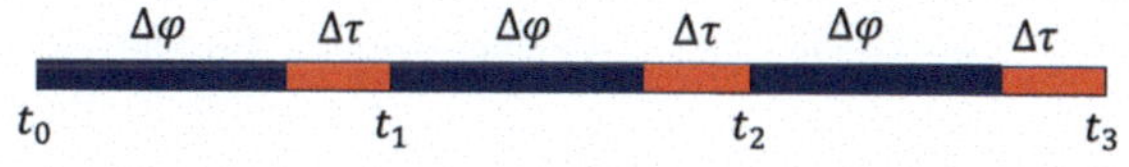

Fig. 5 Set of switching time slots

3.3 Traffic Scheduling

During the switching time slot, if satellite A finds that the meteorological conditions above GST candidates in area 1 are all not suitable for the link construction, then the traffic scheduling will be triggered [9]. This paper adopts a traffic scheduling mechanism, namely, in the switching slot, if there is no GST candidate available in the area, the system adopts the following actions [10].

In a single area of GEO satellite service, special optical cables are constructed between GSTs to achieve low cost, short distance and data sharing. The GSTs between the areas build long-distance temporary optical cable, relatively cost more. In the switching time slot, backbone GEO A will schedule the data generated or relayed onboard to GEO B. Then GEO B will carry out data downlink transmission in area 2 and unload the data to the GST 4 along the satellite-ground link. In other words, before we introduce the mechanism, once GST candidates in area 1 are all inactive, satellite-ground backbone link will be interrupted. However now the significant service which carried by backbone links will not be discarded. For example, a user directly obtained the data unloaded by GEO A from the GST 1, but now the GST 1 needs to obtain the data from the GST 4 through the long-distance optical cable transmission. The transmission path changes to GEO A—GEO B—GST 4—GST 1. Compared with GEO A—GST 1, the number of hops increases by two hops. After this switching time slot, users in area 1 will still not lose service, increasing latency costs accordingly. We can get the delay cost as follows:

$$t_{delay} = \frac{s}{c} + \frac{d_{usr-1}}{c_{fiber}} \tag{2}$$

$$t'_{delay} = \frac{l}{c} + \frac{s}{c} + \frac{d_{1-4} + d_{1-usr}}{c_{fiber}} \tag{3}$$

$$\Delta t_{delay} = t'_{delay} - t_{delay} = \frac{l}{c} + \frac{d_{1-4}}{c_{fiber}} \tag{4}$$

According to the above formula, we can calculate it briefly as follows. t_{delay} represents the average delay before and t'_{delay} represents the average delay after. The height of the synchronous orbit is 36000 km, which is represented as s. The value of l, which represents the communication distance between GEO satellites, is $36000\sqrt{3}$ km according to the geometric relationship. d_{m-n} represents the distance between source m and destination n. The propagation rate of the optical fiber is 2/3 of the speed of light c [2]. Considering the breadth of GEO's coverage area, we take Beijing-San Francisco as an city-pair example. In the experiment 9600 km is taken in a straight line.

By plugging in the values, we get the results as follows:

$$\Delta t_{delay} = 10\,\text{ms}\; t_{delay} = 125\,\text{ms} \tag{5}$$

$$\eta = \frac{\Delta t_{delay}}{t_{delay}} \tag{6}$$

The use of traffic scheduling mechanism will cause an average delay cost of 8%. On average, the loss is on the order of a few percent, which are negligible.

4 Performance Analysis

We construct the satellite-ground laser network experiment system proposed in the previous chapter for simulation. The experiment uses Python to simulate the communication process. Figure 6 compares the differences in network availability between re-routing system and the improved predictable system. As is known to us, the results of prediction can affect the link decision. If we predict that the weather above the GST will still be in good condition at the next moment, but the actual weather condition at the next moment is poor, it will cause an unpredictable interruption. when the accuracy of prediction is 100%, we can get the optimal link decision. Considering the accuracy of the prediction, the discussion is divided into three cases respectively the prediction accuracy is 50, 70 and 100%. In each experiment case, 500 time slots are selected for testing to observe the proportion of effective transmission time slots. The results are shown in Fig. 6.

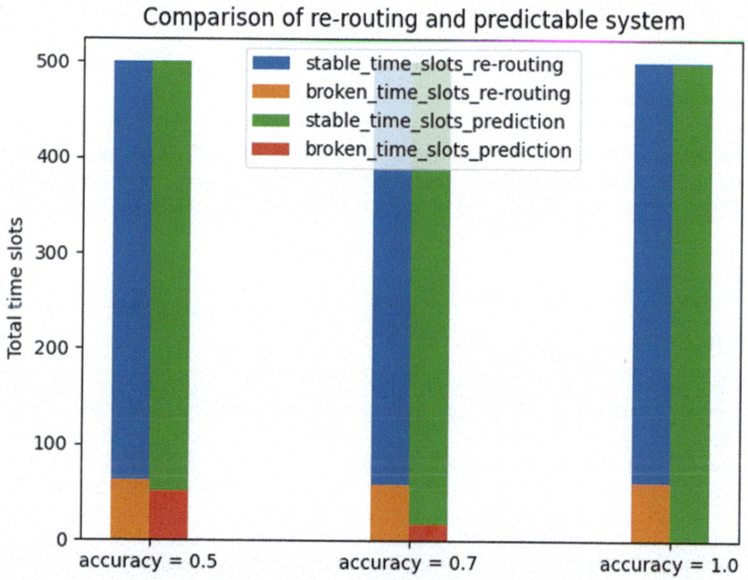

Fig. 6 Comparison of re-routing and predictable system

Obviously, when the prediction accuracy is 100%, the availability of the network should also be the optimal. With the increase of the prediction accuracy, the difference between the prediction network model and the re-routing network model will also grow. The result of the experiment verifies the effectiveness of the link switching mechanism.

5 Conclusion

The traditional GST site diversity method adopts a passive response mechanism. When the link interruption occurs over the GST, the rerouting mechanism will be triggered, and the system will search for the next available GST. During the re-routing period the backbone link will be interrupted so that the high-speed service cannot be maintained, which will have a serious impact on the system. This paper proposes a link switching and traffic scheduling method for satellite-ground uninterrupted communication. Compared with the traditional site diversity mode, the technique actively switch links and perform traffic scheduling based on meteorological prediction. This method improves the availability of network and avoids temporary interruption. It will provide an important guarantee for long-term uninterrupted communication of satellite-ground laser links in the future.

In this paper, our discussion is based on the assumption that one SAT can only communicate with one GST at a time. In this case, the link will be interrupted during switching, which cannot guarantee complete uninterrupted communication. Therefore, in the future, multi-laser terminal satellites can be introduced to avoid link interruption. In addition, it is also possible to discuss the hybrid laser-microwave link, which is worth studying in the future based on this paper.

Acknowledgements This work is supported by the 13th Five-Year Civil Aerospace Technology Pre Research Project, the Fundamental Research Funds for the Central Universities under Grant 021014380187 and the National Natural Sciences Foundation of China under Grant 62131012.

References

1. Majumdar A, Riclkin JC (2007) Free-space laser communications: principles and advances. Springer
2. Giuliari G et al (2020) Internet backbones in space. ACM SIGCOMM Comput Commun Rev 50(1):25–37
3. Shi XJ et al (2015) Convolutional LSTM network: a machine learning approach for precipitation nowcasting. In: Advances in neural information processing systems, vol 28 (Nips 2015)
4. Bertin C et al (2015) Prediction of optical communication link availability: real-time observation of cloud patterns using a ground-based thermal infrared camera. In: Optics in atmospheric propagation and adaptive systems XVIII. 2015. International Society for Optics and Photonics
5. Li Z et al (2018) Internet anycast: performance, problems, & potential. In: Proceedings of the 2018 conference of the ACM special interest group on data communication

6. Lyras N (2020) Optical satellite networks performance: channel modeling, mitigation techniques & optimization
7. Clark GJ et al (2016) Architecture for cognitive networking within NASA's future space communications infrastructure. In: 34th AIAA international communications satellite systems conference
8. Fuchs C, Moll F (2015) Ground station network optimization for space-to-ground optical communication links. J Opt Commun Netw 7(12):1148–1159
9. Liu S et al (2020) Inter-satellite link assignment for the laser/radio hybrid network in navigation satellite systems. GPS Solut 24(2)
10. Williams WD, Collins M, Boroson DM et al (2007) RF and optical communications: a comparison of high data rate returns from deep space in the 2020 timeframe. In: 12th Ka and broadband communications conference

Direction Finding Based on Spatial Sparsity of Cumulative Slicing for Circular Array in Correlated Noise and Multipath

Ying Xiao, Zhen Huang, and Dingli Xue

Abstract The performance of direction finding for uniform circular array (UCA) is degraded with low precision and typical false-alarm result from the correlated noise and coherent interference. To address this issue, a novel direction-finding algorithm based on spatial sparsity representation of high-order cumulative slicing is presented. Spatial sparsity reconstruction based on sparse Bayesian inference of high-order cumulative slicing is constructed to simultaneously mitigate the correlated noise introduced from the environment and the coherent interference caused by multipath. Meanwhile, the multiple interpolation is utilized to overcome the quantization error inherent in sparsity representation. The implementation of high-precision direction-finding is achieved by optimizing each individual component rather than relying on global fitting. The simulation results demonstrate that superior direction-finding accuracy and computational efficiency are provided for UCA, even with concurrent occurrences of environmentally correlated noise and coherent interference.

Keywords Direction finding · High-order cumulative slicing · Spatial sparsity representation · Correlated noise · Coherent interference

---

Y. Xiao
Department of Electronic Engineering, Tsinghua University, Beijing, China
e-mail: xy20@mails.tsinghua.edu.cn

Z. Huang (✉) · D. Xue
Beijing National Research Center for Information Science and Technology, Tsinghua University, Beijing, China
e-mail: huangzhen@tsinghua.edu.cn

D. Xue
e-mail: dinglixue@tsinghua.edu.cn

M. Ma (ed.), *Proceedings of the 12th International Conference on Communications, Circuits, and Systems*, Lecture Notes in Electrical Engineering 1193,
https://doi.org/10.1007/978-981-97-2636-3_6

1 Introduction

Direction of arrival (DOA) estimation is a widely researched field in radio detection [1–4]. Algorithms based on subspace decomposition, such as MUSIC [5] and ESPRIT [6], realize super resolution by transitioning from high resolution. However, these algorithms rely on the covariance matrix decomposition property of signals and assume that the noise is independent. In reality, the ambient noise received by each antenna element is correlated mutually [7–9], and mixed coherent interference is received owing to multipath propagation. Consequently, most subspace-based algorithms work fail in this respect. Regarding the coherent signal issue, pre-processing approaches such as FBSS [10] and Toeplitz matrix reconstruction [11] are employed for decoherence. But they work properly only applying to the Uniform Linear Array (ULA). In the case of a Uniform Circular Array (UCA), it needs to be approximated to a virtual linear array by mode transformation, which has strict requirements for the number of antenna elements. With the limited number, direction-finding accuracy decreases directly because of the high approximation error. Weighted subspace fitting [12, 13] and maximum likelihood algorithms [14] can be used for coherent signals without relying on subspace decomposition, but their computational complexity is considerably high. Currently, sparse-based algorithms have gained widespread attention. Sparse signal reconstruction algorithms, such as L1-SVD [15], reduce sensitivity to coherent interference but suffer from fixed biased. Subsequently, algorithms based on sparse Bayesian inference are developed, including OGSBL [16], PSBL [17], Correlated-VBI [18], which are used to solve the problem of coherent interference in direction finding for UCA. Nevertheless, they are vulnerable to ambient noise, and in situations with a low signal-to-noise ratio (SNR), the precision has severely deteriorated. Meanwhile, most algorithms based on sparse reconstruction prioritize the optimization of the overall fitting error, which conflicts with the emphasis on individual component accuracy in direction finding [19]. Overall, currently available algorithms are incapable of effectively managing correlated ambient noise and coherent interference in direction finding for UCA.

In the paper, a sparse representation algorithm for circular arrays is introduced. The high-order cumulative slicing is constructed to suppress the correlated noise emanating from the environment because of its blind characteristic of the Gaussian process. Additionally, a sparse Bayesian model is established by using the high-order cumulative slicing as the observation quantity to reconstruct the space sparsity of incoming signals, allowing the algorithm to handle the interference which is coherent with the source. Hierarchical interpolation reduces quantization error, to overcome the quantization error inherent in sparsity representation and obtains a more accurate secondary reconstruction result. Finally, based on the reconstructed spatial features, signal components are separated and high-precision direction-finding is performed for each individual.

The present paper is structured in the following manner: Sect. 2 introduces the array manifold for UCA. The signal model with correlated noise and coherent interference is formulated. In Sect. 3, we detail the algorithm's composition and primary steps, as well as analyze the computational complexity. In Sect. 4, simulation experiments are conducted to confirm the effectiveness of the presented method. Section 5 is the conclusion.

Notations: The transpose and conjugate transpose operations are denoted by $(\cdot)^{\mathrm{T}}$, $(\cdot)^{\mathrm{H}}$, respectively. And $(\cdot)^{*}$, $(\cdot)^{-1}$ represent inverse and complex conjugate operations. The bold italic characters symbolize matrices and vectors. $\mathrm{Tr}(\cdot)$ is the trace operator. $|\cdot|\|\cdot\|_2$, $\|\cdot\|_F$ represent determinant, l_2 norm and Frobenious norm, respectively. **0** and **I** are all-zero matrix and the identity matrix of the specified dimensions.

2 Problem Formation

The UCA comprises multiple antenna elements uniformly distributed on a circular ring. Assuming the circle radius is R, and M antenna elements are located in the XOY plane. The first element is placed on the X-axis, with the remaining elements numbered in a counterclockwise direction starting from 2 to *M*. The angle of each element are $\phi_m = 2\pi(m-1)/M$, where $m = 1, 2, \ldots, M$.

For one-dimensional direction finding, *D* narrowband signals are simultaneously incident from different directions $\Theta = \{\theta_1, \theta_2, \ldots, \theta_D\}$ in the far field, where $\theta_i \in [0, 2\pi]$ is the azimuth angle. Taking the circle center as the phase reference point, the array manifold matrix of UCA is represented as $\boldsymbol{A}(\boldsymbol{\Theta}) = [\boldsymbol{a}(\theta_1), \boldsymbol{a}(\theta_2), \ldots, \boldsymbol{a}(\theta_{\mathrm{D}})]$, where $\boldsymbol{a}(\theta_2)$ is the array manifold vector, which is defined as

$$\boldsymbol{a}(\theta_i) = \begin{bmatrix} exp\{j2\pi R/\lambda cos(\theta_i - \phi_1)\} \\ exp\{j2\pi R/\lambda cos(\theta_i - \phi_2)\} \\ \vdots \\ exp\{j2\pi R/\lambda cos(\theta_i - \phi_M)\} \end{bmatrix} \tag{1}$$

The array output of UCA at time t is expressed as

$$\boldsymbol{X}(t) = \boldsymbol{A}(\Theta)\boldsymbol{S}(t) + \boldsymbol{N}(t) \tag{2}$$

where $\boldsymbol{S}(t) \in \mathbb{C}^{D\times L}$ is the matrix of incident signals. Due to the influence of multipath propagation, there exist coherent signals in S(t), such as $s_j(t) = \eta e^{j\Delta\psi} s_i(t)$, where η is the fading factor of the amplitude and $\Delta\psi$ is the phase difference. $\boldsymbol{N}(t) \in \mathbb{C}^{M\times L}$ is the received noise originating from the environment. And the rows within it are correlated [20]. L is the number of samples.

3 The Proposed Algorithm

A. Sparse Bayesian model for High-Order Cumulative Slicing

High-order cumulants have blind characteristics for Gaussian processes and are used to suppress Gaussian noise. Compared with second-order statistics, they provide richer statistical information and can improve the performance of direction-finding algorithms in non-Gaussian incoming signal situations. In this paper, we construct a high-order cumulative slicing of X(t) to suppress the impact of environmental noise. The expression is

$$\boldsymbol{C}_x^{slice} = cum\{\boldsymbol{X}(t), x_1(t), \boldsymbol{X}^{\mathrm{H}}(t), x_1^*(t)\} = A(\Theta)\boldsymbol{C}_x^{slice}\mathbf{A}^{\mathrm{H}}(\Theta) \tag{3}$$

where $\mathrm{C}_{\mathrm{s}}^{\mathrm{slice}}$ is the high-order cumulative slicing of S(t), given by

$$\boldsymbol{C}_x^{slice} = \sum_{d_1=1}^{\mathrm{D}}\sum_{d_2=1}^{\mathrm{D}} a_1(\theta_{d1})a_1^*(\theta_{d2})cum\{\boldsymbol{S}(t),\ s_{d_1}(t),\ \boldsymbol{S}^{\mathrm{H}}(t),\ S_{d_2}^*(t)\} \tag{4}$$

With the properties of high-order cumulant, it's evident that the constructed high-order cumulative slicing contains all the information about the incoming direction of the signals, while eliminating other redundant information and preserving the Gaussian noise blind characteristics.

Without considering quantization error, the full azimuth range 0–360° is divided into F possible angles by an angular interval of $\Delta\vartheta$, forming an over-complete set of angles $\boldsymbol{\Omega} = \{\vartheta_1, \vartheta_2, \ldots, \vartheta_{\mathrm{F}}\}$. F is much larger than the number of the incoming signals D. When $\Delta\vartheta$ is small enough, the DOA of the incoming signals $\boldsymbol{\Theta} = \{\theta_1, \theta_2, \ldots, \theta_{\mathrm{D}}\}$ is a small subset of $\boldsymbol{\Omega}$, i.e. $\boldsymbol{\Theta} \subset \boldsymbol{\Omega}$. Meanwhile, the array output model in (2) is extended to the over-complete array output model, given by

$$\boldsymbol{X}(t) = \boldsymbol{A}(\Omega)\tilde{\mathbf{S}}(t) + \boldsymbol{N}(t) \tag{5}$$

where $\boldsymbol{A}(\boldsymbol{\Omega}) \in \mathbb{C}^{M\times F}$ is the over-complete form of $\boldsymbol{A}(\boldsymbol{\Theta})$. Each column in it is known as a basis function corresponding to the array manifold vector at each element in $\boldsymbol{\Omega}$, as $\mathbf{A}(\boldsymbol{\Omega}) = [a(\vartheta_1), a(\vartheta_2), \ldots, a(\vartheta_{\mathrm{F}})]$.

The complete set of signals $\tilde{\mathbf{S}}(t) \in \mathbb{C}^{F\times L}$ is the zero-padding extension form of incoming signals, defined as

$$\tilde{\mathbf{S}}(t) = \begin{cases} \mathbf{S}(t) & \vartheta_f = \theta_d \\ \mathbf{0}_{1\times L} & others \end{cases} \tag{6}$$

Clearly, $\tilde{\mathbf{S}}(t)$ is a D-rows sparse matrix. In the same way, $\boldsymbol{C}_s^{slice}$ is mapped to the over-complete angle set expansively as $\tilde{\boldsymbol{C}}_s^{slice}$. Considering the high-order cumulative slice estimate error caused by the limited number of snapshots, the observation equation is obtained as

$$C_x^{slice}=A(\Omega)G_S+E \tag{7}$$

where $\boldsymbol{G}_s = \tilde{\boldsymbol{C}}_s^{slice} A^{\mathrm{H}}(\Omega)$, $\boldsymbol{G}_s \in \mathbb{C}^{F\times M}$ is a sparse matrix defined in the higher-order cumulative slice space, equipping the same row sparsity as $\tilde{\mathbf{S}}(t)$.

Then direction finding can be carried out by reconstructing the spatial sparsity of $\boldsymbol{G}_s$. Firstly, a hierarchical prior Bayesian model is established by introducing hyper-parameter $\gamma = [\gamma_1, \gamma_2, \ldots, \gamma_{\mathrm{F}}]^{\mathrm{T}}$ to describe the prior of $\boldsymbol{G}_s$, called equivalent signal strength. Assume that each column of $\boldsymbol{G}_s$ follows a complex Gaussian distribution as

$$\boldsymbol{g}_s(l) \sim CN(\boldsymbol{0}, \boldsymbol{\Gamma}), l = 1, 2, \ldots, M \tag{8}$$

Furthermore, the prior probability density function (PDF) of $\boldsymbol{G}_s$is derived as

$$\mathrm{p}(\boldsymbol{G}_s|\Upsilon) = \prod_{l=1}^{M} CN(\boldsymbol{g}_s(l)|(\boldsymbol{0}, \boldsymbol{\Gamma}) \tag{9}$$

where $\boldsymbol{\Gamma} = diag(\Upsilon)$. Obviously, when $\gamma_f = 0$, it indicates that the entire f-th row of $\boldsymbol{G}_s$ is zero, which means that no signal is incoming at ϑ_f. The hyperprior PDF of Υ is assumed to be

$$p(\Upsilon) = \prod_{f=1}^{F} \mathrm{Gamma}(\gamma_f; 1, \rho) \tag{10}$$

where $Gamma(x; a, b)$ means the variable x follows a Gamma distribution with parameters (a,b). When $a = 1$, it's equivalent to an exponential distribution with parameter b. To better control the spatial sparsity of the equivalent signal strength, ρ is set closely to 0. Then we get $\mathrm{p}(\boldsymbol{G}_s, \Upsilon) = \mathrm{p}(\boldsymbol{G}_s|\Upsilon)\mathrm{p}(\Upsilon)$.

Similarly, the prior distribution of the observed noise is given by

$$\mathrm{p}(\boldsymbol{E}|\alpha_0) = \prod_{l=1}^{M} CN(e(l)|\boldsymbol{0}, \alpha_0^{-1}\boldsymbol{I}_M) \tag{11}$$

where $\boldsymbol{e}(l) \sim CN(\boldsymbol{0}, \alpha_0^{-1}\boldsymbol{I}_M)l = 1, 2, \ldots, M$. And α_0 is defined as a random variable called noise accuracy, and its hyperprior distribution is expressed as

$$\mathrm{p}(\alpha_0) = \mathrm{Gamma}(\alpha_0; 1, b) \tag{12}$$

where b is a minuscule positive parameter. Therefore, $\boldsymbol{C}_s^{slice}$ follows complex Gaussian distribution as

$$p(\boldsymbol{C}_x^{Slice}|G_s; \alpha_0) = CN(\tilde{\boldsymbol{A}}\boldsymbol{G}_s, \alpha_0^{-1}\boldsymbol{I}_M) \tag{13}$$

Combining the hierarchical prior structure, the joint PDF is

$$p(\boldsymbol{G}_s, \boldsymbol{C}_x^{slice}, \Upsilon, \alpha_0) = p\left(\boldsymbol{C}_x^{slice} | \boldsymbol{G}_s; \alpha_0\right) p(\boldsymbol{G}_s | \Upsilon) p(\Upsilon) p(\alpha_0) \tag{14}$$

The posterior probability density function of the $\boldsymbol{G}_s$ with respect to the observed quantity $\boldsymbol{C}_s^{slice}$ is obtained by Bayesian theory,

$$p(\boldsymbol{G}_s | \boldsymbol{C}_x^{slice}; \Upsilon, \alpha_0) = \frac{p(\boldsymbol{C}_x^{slice} | \boldsymbol{G}_s; \alpha_0) p(\boldsymbol{G}_s | \Upsilon)}{p(\boldsymbol{C}_x^{slice}; \Upsilon, \alpha_0)} = CN\left(\boldsymbol{\mu}_s, \sum\nolimits_s\right) \tag{15}$$

where $\boldsymbol{\mu}_s$, $\sum_s$ are given as follows

$$\boldsymbol{\mu}_s = \boldsymbol{\Gamma} \tilde{\mathrm{A}}^{\mathrm{H}} \sum\nolimits_x^{-1} \boldsymbol{C}_x^{slice} \tag{16}$$

$$\sum\nolimits_s = \left(\alpha_0 \tilde{\boldsymbol{A}}^{\mathrm{H}} \tilde{\boldsymbol{A}} + \boldsymbol{\Gamma}^{-1}\right)^{-1} = \boldsymbol{\Gamma} - \boldsymbol{\Gamma} \tilde{A}^{\mathrm{H}} \sum\nolimits_x^{-1} \tilde{A} \sum\nolimits_x^{-1} \tilde{A} \boldsymbol{\Gamma} \tag{17}$$

where, $\sum_x = \alpha_0^{-1} \boldsymbol{I}_M + \tilde{A} \boldsymbol{\Gamma} \tilde{A}^{\mathrm{H}}$ is the covariance matrix of $\boldsymbol{C}_x^{slice}$.

B. Spatial sparse feature reconstruction

Based on the analysis in the previous section, the sparse representation of high-order cumulative slicing on the angle-complete set $\boldsymbol{\Omega}$ can be achieved by solving the maximum value of (15). The objective function is expressed as:

$$\widehat{\boldsymbol{G}}_s = \max_{\boldsymbol{G}_s} \left\{ p(\boldsymbol{G}_s | \boldsymbol{C}_x^{slice}; \widehat{\Upsilon}, \widehat{\alpha}_0) \right\} \tag{18}$$

The hyperparameter Υ determines the spatial strength prior distribution of $\boldsymbol{G}_s$ α_0 is used to characterize the intensity of observation noise and control the fitting error. Therefore, it's necessary to optimize the hyperparameters (Υ, α_0) using the observed data to obtain their estimated values $(\widehat{\Upsilon}, \widehat{\alpha}_0)$. The objective function for optimization is

$$\left(\widehat{\Upsilon}, \widehat{\alpha}_0\right) = \arg\max_{\Upsilon, \alpha_0} \{p(\Upsilon, \alpha_0 | \boldsymbol{C}_x^{slice})\} \tag{19}$$

Owing to $p(\Upsilon, \alpha_0 | \boldsymbol{C}_x^{slice}) p(\boldsymbol{C}_x^{slice}) = p(\boldsymbol{C}_x^{slice}; \Upsilon, \alpha_0)$, the objective function in (19) is equivalent to

$$\left(\widehat{\Upsilon}, \widehat{\alpha}_0\right) = \arg\max_{\Upsilon \geq 0, \alpha_0 > 0} \{p(\boldsymbol{C}_x^{slice}; \Upsilon, \alpha_0)\} \tag{20}$$

Further optimization of the objective function can be achieved by taking the logarithm and removing the constant term, resulting in.

$$l(\gamma, \alpha_0) = -M\ \ln\left|\sum\nolimits_x\right| - \mathrm{Tr}\left[\left(\boldsymbol{C}_x^{slice}\right)^H \sum\nolimits_x^{1} \boldsymbol{C}_x^{slice}\right] + \sum_{f=1}^{F}(-\rho\gamma_f) - b\alpha_0 \quad (21)$$

Due to the high dimensionality of the parameters to be optimized and the nonlinearity of the objective function, the optimization process is implemented using EM algorithm. First is the E-step, which calculates the mean and covariance matrix of $\boldsymbol{G}_s$ using (16) and (17). Then, in the M-step, the iterative formula for $l\ (\Upsilon, \alpha_0)$is obtained by solving the partial derivative of (21) with respect to optimizable parameters, given by

$$\gamma_f^{(new)} = \frac{\sqrt{M^2\varpi_f^2 + 4\rho\left\|\boldsymbol{\mu}_{s(f,:)}\right\|_2^2} + M\varpi_f}{2\rho}, \quad f = 1, 2, \ldots, F \quad (22)$$

$$\alpha_0^{(new)} = \frac{M^2 - M\alpha_0^{(last)} Tr\left(\tilde{\mathrm{A}}\Gamma\tilde{\mathrm{A}}^{\mathrm{H}} - (\tilde{\mathrm{A}}\Gamma\tilde{\mathrm{A}}^{\mathrm{H}})\sum\nolimits_x^{-1}\left(\tilde{\mathrm{A}}\Gamma\tilde{\mathrm{A}}^{\mathrm{H}}\right)\right)}{\left\|\boldsymbol{C}_x^{slice} - \tilde{\mathrm{A}}\boldsymbol{G}_s\right\|_{\mathrm{F}}^2 + b} \quad (23)$$

At the same time, the noise variance is updated as $\sigma^{2(new)} = 1/\alpha_0^{(new)}$.

In the M-step, the iterative solution process only uses the diagonal elements of $\sum_s$. To reduce the computational complexity, (17) is replaced by

$$\sum\nolimits_{s(f,f)} = \gamma_f - \gamma_f^2 \boldsymbol{W}_{(f,:)}\tilde{\boldsymbol{A}}_{(:,f)} \quad (24)$$

Through the EM algorithm with (16), (22)–(24), the spatially sparse feature reconstruction of high-order cumulative slicing is accomplished.

C. High precision direction finding based on spatial sparse feature

DOA estimation is essentially a continuous parameter problem, so discrete angle grids cause unavoidable quantization errors. In this paper, the multi-interpolation method is used to reduce the quantization errors of reconstruction in off-grid situations.

Firstly, according to the reconstruction result of the spatially sparse feature $\widehat{\Upsilon}$, the first M–1 largest spectral peaks are selected. Setting a normalized threshold and selecting peaks higher than the threshold, the corresponding angles are called the initial estimate results $\boldsymbol{\Theta}_{pre} = \left\{\overline{\vartheta}_1, \overline{\vartheta}_2, \ldots, \overline{\vartheta}_P\right\}$ and $\boldsymbol{\Theta}_{pre} \subset \boldsymbol{\Omega}$. When the actual arrival direction θ_p of the p-th signal does not fall exactly on the angle grid point, i.e. $\theta_p \neq \overline{\vartheta}_p$, there is a sub-peak $\gamma_{\overline{\vartheta}p} = \max\left\{\gamma_{\overline{\vartheta}p+\Delta\vartheta}, \gamma_{\overline{\vartheta}p-\Delta\vartheta}\right\}$ near the peak value $\gamma_{\overline{\vartheta}p}$. The angle grid points are $\overline{\vartheta}_p$ and $\overline{\vartheta}'_p$correspondingly. Assuming that $\overline{\vartheta}'_p < \overline{\vartheta}_p$, we get $\overline{\vartheta}'_p < \theta_p < \overline{\vartheta}_p$. Therefore, the actual direction of the signal is approximately estimated using linear interpolation, described as

$$\tilde{\theta}_p = \overline{\vartheta}'_p + \frac{\sqrt{\Upsilon_{\overline{\vartheta}'_p}}}{\sqrt{\Upsilon_{\overline{\vartheta}'_p}} + \sqrt{\Upsilon_{\overline{\vartheta}_p}}} \overline{\vartheta}_p, \; p = 1, 2, \ldots, P \tag{25}$$

The interpolated approximate estimation results are denoted as $\boldsymbol{\Theta}_{in} = \left\{\tilde{\theta}_1, \tilde{\theta}_2, \ldots, \tilde{\theta}_{\mathrm{P}}\right\}$. Then, bilinear interpolation is performed between the rough estimate results and the angle grid points. A middle angle value is inserted on both sides to form the bilinear interpolation result as

$$\tilde{\theta}'_p = \tilde{\theta}_p - \frac{\tilde{\theta}_p - \tilde{\theta}'_p}{2}, \; p = 1, 2, \ldots, P \tag{26}$$

$$\tilde{\theta}''_p = \tilde{\theta}_p - \frac{\tilde{\theta}_p - \tilde{\theta}'_p}{2}, \; p = 1, 2, \ldots, P \tag{27}$$

Meanwhile, the candidate angle set is obtained as

$$\boldsymbol{\Omega}_{new} = \boldsymbol{\Theta}_{pre} \cup \boldsymbol{\Theta}_{in} = \{\xi_1, \xi_2, \ldots, \xi_{\overline{F}}\} \tag{28}$$

Where $\boldsymbol{\Theta}_{in} = \left\{\tilde{\theta}_1, \tilde{\theta}_2, \ldots, \tilde{\theta}_{\mathrm{P}}\right\} \cup \left\{\tilde{\theta}'_1, \tilde{\theta}'_2, \ldots, \tilde{\theta}'_{\mathrm{P}}\right\} \cup \left\{\tilde{\theta}''_1, \tilde{\theta}''_2, \ldots, \tilde{\theta}''_{\mathrm{P}}\right\}$ and $\boldsymbol{\Theta}_{pre} = \{\overline{\vartheta}_1, \overline{\vartheta}_2, \ldots, \overline{\vartheta}_P\} \cup \left\{\overline{\vartheta}'_1, \overline{\vartheta}'_2, \ldots, \overline{\vartheta}'_P\right\}$. $\overline{F}$ is the number of elements in $\boldsymbol{\Omega}_{new}$.

To enhance the accuracy of reconstructing spatial sparse features, a secondary estimation of the equivalent signal strength Υ is performed. Constructing an overcomplete manifold matrix $\overline{\mathbf{A}} = \mathbf{A}(\boldsymbol{\Omega}_{new})$ according to $\boldsymbol{\Omega}_{new}$, and then iteratively compute (16), (22)–(24) to get secondary reconstruction result $\overline{\Upsilon}$. The covariance matrices of $\boldsymbol{C}_x^{slice}$ and $\boldsymbol{G}_s$ are obtained as $\overline{\sum}_x$ and $\overline{\boldsymbol{\Gamma}}$.

With a reliable threshold, the largest D peaks of $\overline{\Upsilon}$ are obtained, and the corresponding angle grids are $\{\overline{\xi}_1, \overline{\xi}_2, \ldots, \overline{\xi}_{\mathrm{D}}\}$. As analyzed before, each spectral peak contains the maximum and the next greatest intensity value, and the angle grids are denoted accordingly as $\{\overline{\xi}_{d1}, \overline{\xi}_{d2}\}$. The high-precision direction-finding range based on $\overline{\Upsilon}$ is given by

$$\tilde{\boldsymbol{\Theta}}_D = \bigcup_{d=1}^{D} \left(\overline{\xi}_{d_1}, \overline{\xi}_{d_2}\right) \tag{29}$$

where $\overline{\xi}_{d1} < \overline{\xi}_{d2}$ is supposed.

Then, high-precision direction finding based on spatial sparse feature is carried out for every single component of incident signals. First, $\{\overline{\xi}_{d1}, \overline{\xi}_{d2}\}$ is removed from $\tilde{\boldsymbol{\Theta}}_D$ to obtain the set $\boldsymbol{A}(\boldsymbol{\Omega}_{new,-d})$ and the corresponding overcomplete array manifold matrix $\overline{\boldsymbol{A}}_{-d} = \boldsymbol{A}(\boldsymbol{\Omega}_{new,-d})$. As well, the covariance matrix of observation is expressed as

$$\overline{\sum_{x}} = \overline{\sum_{x,-d}} + \breve{\gamma}_d a(\theta_d) a^{\mathrm{H}}(\theta_d) \tag{30}$$

where $\overline{\sum}_{x,d} = \overline{\sigma}^2 \boldsymbol{I}_M + \overline{\boldsymbol{A}}_{-d} \overline{\boldsymbol{\Gamma}}_{-d} \overline{\boldsymbol{A}}^{\mathrm{H}}_{-d}$ is a partial form of $\overline{\sum}_{x}$. θ_d, $\breve{\gamma}_d$ are respectively the true DOA and power of the target signal component. Combined with (21), the cost function of every single component is obtained as

$$l(\theta_d, \breve{\gamma}_d) = M \ln(1 + \breve{\gamma}_d p_d) + \frac{q_d}{\breve{\gamma}_d^{-1} + p_d} - \rho \breve{\gamma}_d \tag{31}$$

where

$$p_d = a^H(\theta_d) \overline{\sum_{x,d}}^{-1} a(\theta_d) \tag{32}$$

$$q_d = a^H(\theta_d) \overline{\sum_{x,d}}^{-1} [\boldsymbol{C}_x^{slice} (\boldsymbol{C}_x^{slice})^H] \overline{\sum_{x,d}}^{-1} a(\theta_d) \tag{33}$$

Taking the partial derivative of (31) with respect to $\breve{\gamma}_d$ and letting it be zero, the estimation of $\breve{\gamma}_d$ is formulated as

$$\breve{\gamma}_d = \frac{-(Mp_d + 2\rho) + \sqrt{(Mp_d + 2\rho)^2 - 4\rho(Mp_d + \rho - q_d)}}{2p_d\,\rho} \tag{34}$$

Then (31) is simplified to estimate θ_d as

$$l(\theta_d) = -M \ln(1 + \breve{\gamma}_d p_d) + \frac{q_d}{\breve{\gamma}_d^{-1} p_d} - \rho \breve{\gamma}_d \tag{35}$$

Regarding the direction finding of the target signal component, it's transformed into finding the maximum value of (35) within the high-precision direction finding range of the component as the following function,

$$\widehat{\theta}_d = \underset{\theta_d \in [\xi_{d1}, \xi_{d2}]}{\arg\max} \{l(\theta_d)\}, d = 1, 2, \ldots, D \tag{36}$$

Within $\tilde{\boldsymbol{\Theta}}_D$, searching uniformly with small steps for the angle corresponding to the maximum spectral peak of (36) an obtain the final high-precision direction finding results.

The proposed algorithm is summarized by the following key steps.:

Algorithm.1.

Input: High-order cumulative slicing of array output $\boldsymbol{C}_x^{slice}$.
Output: DOA $\boldsymbol{\Theta}=\{\theta_1,\theta_2,\cdots,\theta_D\}$
Part1. *Spatial sparsity reconstruction*
- Initialization: $\Upsilon^{(0)}$, $\alpha_0^{(0)}$
- While *not converge* do
- Update Υ, α_0 by using (16), (22)-(24)
- End while

Part2. *Multiple interpolation and quadratic reconstruction*
- Multiple interpolation in (25)-(27)
- Initialization: $\overline{\Upsilon}^{(0)}$
- While *not converge* do
- Update $\overline{\Upsilon}$ by using (16), (22), (24)
- End while

Part3. *Signal separation and high precision direction finding*
- Obtain $\tilde{\boldsymbol{\Theta}}_D$ in (29)
- Separate the target signal component in (30)
- Accomplish DOA estimation via (36)

Return DOA results

D. Computational complexity analysis

According to the structure of the proposed algorithm, the total computational complexity is divided into three parts to calculate. Usually we have $D < M < \overline{F} \ll F$. In Part 1, the complexity is determined by both the amount of calculation in a single iteration and the number of iterations. In a single iteration, it takes $O(3M^2F+M^3)$ to obtain $\boldsymbol{\mu}_s$, and O((M + 1)MF) related to $\sum_{s(f,f)}$. Meanwhile, the complexity to update Υ, α_0 is O(MF) and $O(M^2F)$ respectively. Ignoring constants and lower-order terms, the complexity of a single iteration is obtained as $O(M^2F)$. So the complexity in Part 1 is$O(i_1M^2F)$, where i_1 is assumed as the number of iterations. Similarly, the complexity of Part 2 is $O(i_2M^2\overline{F})$ as i_2 is the iterations. In Part 3, the complexity is $O(M^2DF_0)$ on the assumption that the search point for a single component is F_0. In all, the total complexity is $O(i_1M^2F+i_2M^2\overline{F}+M^2DF_0)$.

4 Simulation Results

In the section, simulations are conducted to demonstrate the superior performance of the presented method in comparison to previously proposed approaches, including L1-SVD in [15], OGSBL in [16], PSBL in [17], Correlate-VBI in [18] SAMV2 and SAMV2-SML in [21]. We assume that the 7-element uniform circular array receives BPSK signals, with the radius of the array being half wavelength. We set $\rho = 0.01,\ b = 10^{-4}$. The iteration threshold is 0.001 and the maximum of iterations is 500 for all methods. In the subsequent simulations, we calculate the root mean squared error (RMSE) and probability of detection (PD) to evaluate the performance. The expressions for them are provided below.

$$\text{RMSE} = \sqrt{\frac{1}{N_t D}\sum_{n=1}^{N_t}\sum_{i=1}^{D}(\widehat{\theta}_{i,n}\theta_i)^2} \tag{37}$$

$$\text{PD} = \frac{N_e}{N_t} \times 100\% \tag{38}$$

where N_t represents the Monte Carlo trial count in our examples, which is equal to 500. And D represents the signal count. In the *n*-th Monte Carlo trial, $\widehat{\theta}_{i,n}$ indicates the estimation for θ_i. In our simulations, we consider $\left|\left(\widehat{\theta}_{i,n} - \theta_i\right)\right| \leq 1.5°$ to be successfully resolved. The number of successful resolutions is denoted by N_e.

E. Comparison of normalized spatial spectrum between different methods

In the first example, we experiment with SNR = 0 dB and L is 512 represents the sample count, comparing the normalized spectral peaks of the proposed algorithm versus five others. A signal and a coherent interference are under consideration. They are incident on the array at azimuth angles −40.6°, 20.3°, respectively. The environmental noise received by each antenna element is correlated, with a correlation coefficient of 0.8. As demonstrated in Fig. 1, our proposed algorithm accurately distinguishes two incident signals, and the spectral peak is positioned precisely. That means the proposed method overcomes the effects of correlated noise and coherent interference. However, the L1-SVD algorithm demonstrates signal loss, while the other four algorithms exhibit varying degrees of false alarms.

F. Comparison of RMSE and PD between different methods

Experiment 2 involves comparing the RMSE and probability of detection between different algorithms. The other conditions of the experiment remain consistent with that of the previous one. Figure 2 illustrates the RMSE of different algorithms as the SNR changes from −10 dB to 10 dB. It's evident that the proposed algorithm has the highest DOA estimation accuracy and is the closest to the CRLB. At the SNR of −10 dB, the RMSE of the proposed algorithm approaches 1°, which is

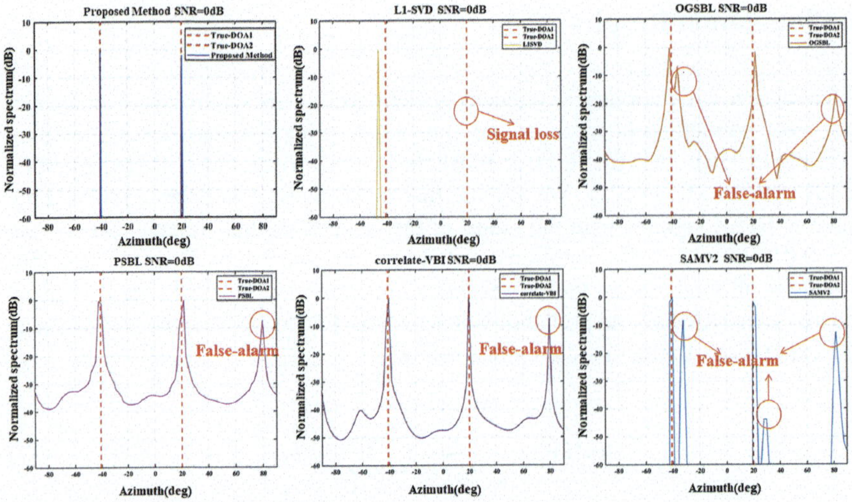

Fig. 1 Normalized spectral peaks of different methods with SNR = 0 dB, L = 512

about 25% of the best algorithm among the other several algorithms. Figure 3 displays the detection probability of each algorithm across varying SNR. When the SNR is −10 dB, the detection probability of the proposed algorithm exceeds 80%, significantly higher than other algorithms. In this case, the detection ability of other algorithms is essentially lost.

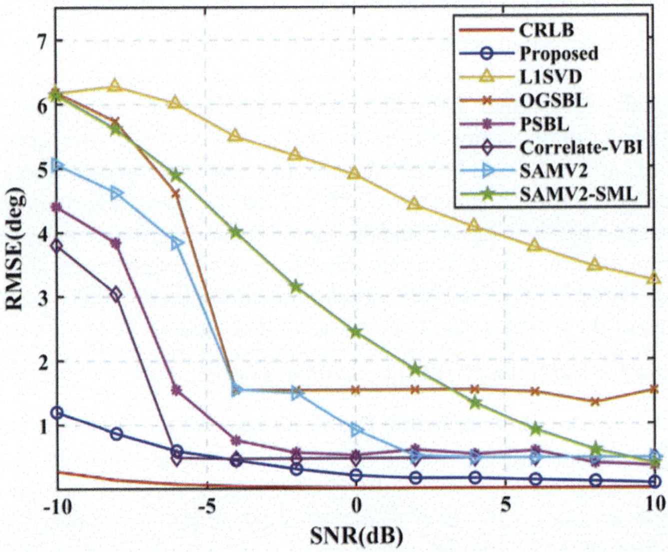

Fig. 2 Comparison of RMSE under various SNRs for different methods with L = 512

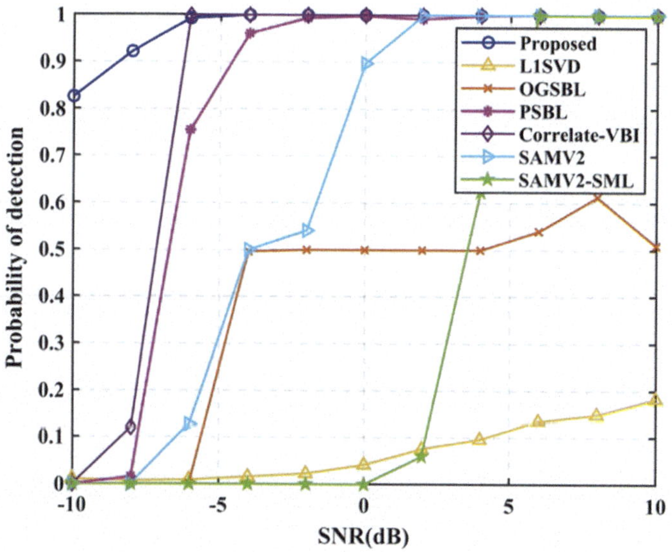

Fig. 3 Comparison of PD under various SNRs for different methods with L = 512

Figures 4 and 5 illustrate the performance of different algorithms as the sample size varies from 100 to 1000. The SNR is set to −5 dB. The algorithm proposed in this paper outperforms competing algorithms at varying sample counts and is unaffected by the number of samples. Even in situations with small samples, high accuracy and detection probability can be achieved. In contrast, the detection probability of L1SVD and OGSBL deteriorates significantly.

G. Comparison of CPU time between different methods

Finally, the CPU time of different methods is compared in the experiment. The algorithms are implemented in Matlab 2020a with Intel(R) Core(TM) i7-10700 K, CPU@3.80 GHz. Tables 1 and 2 respectively show the changes of CPU time with SNR and sample numbers. The proposed algorithm, SAMV2, and SAMV2-SML have demonstrated high speeds, with SAMV2 being the fastest. The CPU time of the proposed algorithm is nearly equal to that of SAMV2. However, the CPU time of L1SVD and PSBL is relatively long. Furthermore, matrix dimension involved in the calculation of the proposed algorithm is independent of the number of sampling points, but the CPU time is increasing with smaller samples because the estimation error of high order cumulative slicing becomes larger, resulting in slower convergence rate. In contrast, the running time of PSBL, OGSBL, and Correlated-VBI substantially increases as the number of samples increases with much samples.

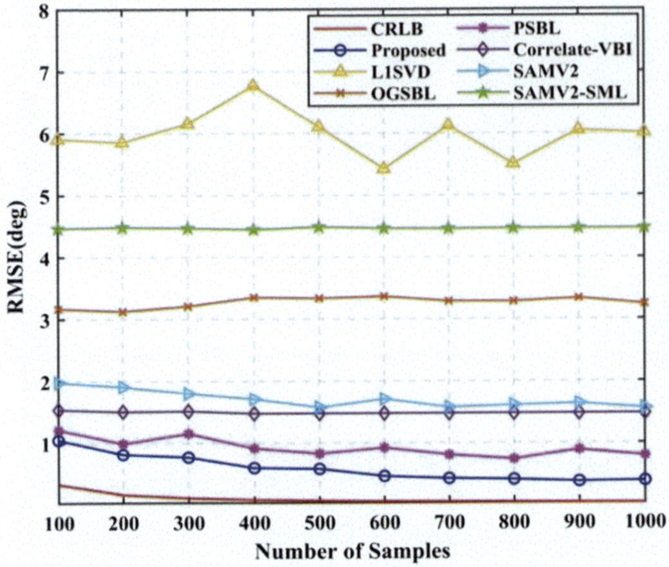

Fig. 4 Comparison of RMSE under various sample numbers for different methods with SNR = −5 dB

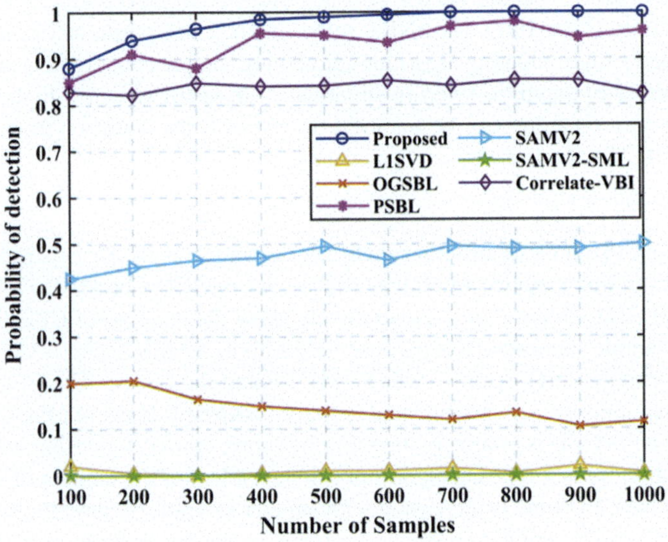

Fig. 5 Comparison of PD under various sample numbers for different methods with SNR = −5 dB

Table 1 Comparison of CPU time (sec) with different SNR

Method	SNR(L = 512)				
	−10 dB	−5 dB	0 dB	5 dB	10 dB
L1-SVD	2.0949	2.0947	2.0970	2.0946	2.0952
OGSBL	1.1374	1.2662	0.9755	0.7594	0.1458
PSBL	1.7399	1.7393	1.5388	1.1701	0.8147
Correlate-VBI	1.2167	1.2173	1.2163	1.2171	1.2155
SAMV2	0.0187	0.0171	0.0126	0.0114	0.0112
SAMV2-SML	0.0510	0.0499	0.0339	0.0304	0.0247
Proposed	**0.0318**	**0.0232**	**0.0186**	**0.0179**	**0.0204**

Table 2 Comparison of CPU time (sec) with different samples

Method	The number of sample(SNR = −5 dB)				
	100	300	500	700	900
L1-SVD	2.4283	2.4185	2.4281	2.4246	2.4299
OGSBL	0.3459	0.5210	0.7347	0.9980	1.1405
PSBL	0.9073	1.2042	2.0477	2.5822	3.2018
Correlate-VBI	0.4218	0.5149	1.4243	1.9055	2.4478
SAMV2	0.0275	0.0250	0.0279	0.0286	0.0280
SAMV2-SML	0.0579	0.0553	0.0589	0.0579	0.0588
Proposed	**0.0330**	**0.0262**	**0.0244**	**0.0241**	**0.0216**

5 Conclusion

The paper presents a novel direction-finding algorithm for UCA, which reconstructs sparsity in space and accomplishes high-precision direction finding based on high-order cumulative slicing. Experimental results demonstrate that the proposed method effectively manages environmental noise and coherent interference in low-SNR situations, compared with previous algorithms. When SNR is −10 dB, the proposed algorithm has a quarter of the RMSE of comparative algorithms, and the correct detection probability is higher than 80%. In contrast, other algorithms are susceptible to false detections which lead to algorithm failure. Additionally, the computational complexity of this algorithm is independent of the number of samples, and the runtime is very close to that of the optimal algorithm in comparison.

References

1. Bekkerman I, Tabrikian J (2006) Target detection and localization using mimo radars and sonars. IEEE Trans Signal Process 54(10):3873–3883
2. Shi J, Wen F, Liu T (2021) Nested mimo radar: Coarrays, tensor modeling, and angle estimation. IEEE Trans Aerosp Electron Syst 57(1):573–585
3. Mao X, Pan H (2013) An improved DOA estimation algorithm based on wavelet operator. J Commun 8(12):839–844
4. Wang L, Chen Z, Wang G (2014) Direction finding and positioning algorithm with COLD-ULA based on quaternion theory. J Commun 9(10):778–784
5. Schmidt R (1986) Multiple emitter location and signal parameter estimation. IEEE Trans Antennas Propag 34(3):276–280
6. Roy R, Kailath T (1989) Esprit-estimation of signal parameters via rotational invariance techniques. IEEE Trans Acoust Speech Signal Process 37(7):984–995
7. Agrawal M, Prasad S (2000) A modified likelihood function approach to DOA estimation in the presence of unknown spatially correlated Gaussian noise using a uniform linear array. IEEE Trans Signal Process 48(10):2743–2749
8. Tan KC, Oh GL (1997) Estimating directions-of-arrival of coherent signals in unknown correlated noise via spatial smoothing. IEEE Trans Signal Process 45(4):1087–1091
9. Wong KM, Reilly JP, Wu Q, Qiao S (1992) Estimation of the directions of arrival of signals in unknown correlated noise. I. The MAP approach and its implementation. IEEE Trans Signal Process 40(8):2007–2017
10. Pillai SU, Kwon BH (1989) Forward/backward spatial smoothing techniques for coherent signal identification. IEEE Trans Acoust Speech Signal Process 37(1):8–15
11. Han F-M, Zhang X-D (2005) An esprit-like algorithm for coherent doa estimation. IEEE Antennas Wirel Propag Lett 4:443–446
12. Viberg M, Ottersten B, Kailath T (1991) Detection and estimation in sensor arrays using weighted subspace fitting. IEEE Trans Signal Process 39(11):2436–2449
13. Viberg M, Ottersten B (1991) Sensor array processing based on subspace fitting. IEEE Trans Signal Process 39(5):1110–1121
14. Stoica P, Sharman K (1990) Maximum likelihood methods for direction-of-arrival estimation. IEEE Trans Acoust Speech Signal Process 38(7):1132–1143
15. Malioutov D, Cetin M, Willsky AS (2005) A sparse signal reconstruction perspective for source localization with sensor arrays. IEEE Trans Signal Process 53(8):3010–3022
16. Yang Z, Xie L, Zhang C (2013) Off-grid direction of arrival estimation using sparse Bayesian inference. IEEE Trans Signal Process 61(1):38–43
17. Wu X, Zhu W-P, Yan J (2016) Direction of arrival estimation for off-grid signals based on sparse bayesian learning. IEEE Sens J 16(7):2004–2016
18. Cao Z, Dai J, Xu W, Chang C (2021) Fast variational bayesian inference for temporally correlated sparse signal recovery. IEEE Signal Process Lett 28:214–218
19. Liu ZM, Huang ZT, Zhou YY (2011) Direction-of-arrival estimation of wideband signals via covariance matrix sparse representation. IEEE Trans Signal Process 59(9):4256–4270
20. Viberg M, Stoica P, Ottersten B (1997) Maximum likelihood array processing in spatially correlated noise fields using parameterized signals. IEEE Trans Signal Process 45(4):996–1004
21. Abeida H, Zhang Q, Li J, Merabtine N (2013) Iterative sparse asymptotic minimum variance based approaches for array processing. IEEE Trans Signal Process 61(4):933–944. (15 Feb 2013)

Next Generation Communication Network Technology and Future Development

Spatial Dynamic Preamble Soft Allocation for mMTC in Massive MIMO Systems

Xiao Fu, Xiaofeng Liu, Xinrui Gong, Rui Sun, Qingguo Shen, and Xiqi Gao

Abstract Massive machine type communications (mMTC) is a key enabler to support massive access of Internet of Things (IoT)in 5G/6G wireless networks. In this paper, a novel spatial dynamic preamble soft allocation scheme is proposed to improve access performance for massive machine-type communications. In the proposed scheme, preambles can be spatially reused due to the plenty spatial degrees of freedom offered by massive multiple-input-multiple-output (MIMO). Moreover, the preambles are dynamically allocated and statistically shared among machine type devices (MTDs) based on the various traffic, which further improves access performance and fairness among MTDs. Simulation results verify the performance advantages of the proposed scheme.

Keywords Machine type communications · Random access · Preamble soft allocation · Massive MIMO

X. Fu (✉) · X. Liu · X. Gong · R. Sun · Q. Shen · X. Gao
National Mobile Communications Research Laboratory, Southeast University, Nanjing 210096, China
e-mail: fu_xiao@seu.edu.cn

X. Liu
e-mail: xf_liu@seu.edu.cn

X. Gong
e-mail: xinruigong@seu.edu.cn

R. Sun
e-mail: ruisun@seu.edu.cn

Q. Shen
e-mail: shenqingguo@pmlabs.com.cn

X. Gao
e-mail: xqgao@seu.edu.cn

X. Fu · X. Liu · X. Gong · Q. Shen · X. Gao
Purple Mountain Laboratories, Nanjing 211100, China

M. Ma (ed.), *Proceedings of the 12th International Conference on Communications, Circuits, and Systems*, Lecture Notes in Electrical Engineering 1193,
https://doi.org/10.1007/978-981-97-2636-3_7

1 Introduction

Massive machine type communications (mMTC) has been one of the significant technology of the fifth-generation/sixth generation (5G/6G) wireless networks [1], which supports the widespread applications of the Internet of Things (IoT), such as smart city, e-health and so on [2]. The characteristic features of machine-type devices (MTDs) are small packet size and tremendous device quantity. Currently, the number of MTDs in the world has explosive growth and is predicted to reach 1.47 billion in 2023 [3].

Differ from human-to-human (H2H) communications with high rate, MTDs usually sporadically transmit small data packets. Each MTD performs a random access (RA) procedure to access the network before data transmission. In RA procedure, each MTD randomly chooses one preamble from the available preamble set and transmit to the base station (BS). If multiple MTDs choose the same preamble (called preamble collision), these collided MTDs will fail in their RA procedures. Due to the limited preamble resources, severe preamble collisions occur when massive MTDs simultaneously access the networks, which significantly degrades access performance.

To improve access performance, several methods have been proposed to mitigate preamble collision. The access class barring (ACB) [4] specified by 3rd generation partnership project (3GPP) is an efficient method in time domain. In ACB schemes, the backlogged MTDs (i.e., newly-wake up MTDs and retransmission MTDs) firstly perform a ACB check, and only the MTDs pass the ACB checks can send their preambles. Other MTDs with failure in ACB checks are barred and need to back off temporarily. The number of MTDs that simultaneously access is controlled and dispersed to a longer period via ACB, which can decrease the preamble collision probability. Further, dynamic ACB schemes (e.g., [5]) have been proposed to achieve better access performance. However, ACB control will cause longer access delay.

Another method is dividing the MTDs into a number of clusters, and each cluster is assigned a dedicated cluster head (CH) device [6, 7]. Clustered MTDs transmit data to its CH instead of the BS, and CH aggregates data and accesses the BS via one dedicated candidate preamble. However, these two-stage access schemes need extra device cost, and the access delay of MTDs increases due to waiting delay of data aggregation.

MTDs can also be grouped based on spatial parameters for collision resolution or preamble reuse. Due to different propagation distance to the BS, MTDs can be distinguished based on their propagation delays. Kim et al. [8] proposed to utilize timing advance (TA) information to assist RA. However, the performance of TA-assisted RA schemes are limited by the size of a cell. In a small cell, the propagation delays among MTDs are close and indistinguishable. In 5G/6G, massive multiple-input multiple-output (MIMO) brings plenty of spatial degrees of freedom in angle of arrival (AOA) domain. He and Ren [9] proposed an AOA based grouped RA scheme with preamble reuse, which effectively improves the RA performance. Specifically, the available preamble set is equally divided into two subsets, and these two subsets are reused every two adjacent groups. Each group is allocated with different subset from its adjacent groups so that MTDs will not select the same preamble with MTDs in adjacent groups, which remove the effects of multipath spread among groups. However, MTDs in each group are allocated with the preamble subset instead of the total set, where their available preamble number is reduced by half. For groups that backlogged MTDs are concentrated in, preamble collision can be more severe due to fewer preambles, which is unfair to MTDs in these groups. To further improve access performance and fairness among groups, we propose a spatial dynamic preamble soft allocation (PSA) scheme, where the preamble subsets are statically shared among groups and the effects of multipath can also be sufficiently mitigated. Simulation results show that the proposed scheme can significantly success RA probability and reduce access delay.

The rest of this paper is organized as follows. Section 2 describes the system model. The proposed spatial dynamic preamble soft allocation scheme is detailed in Sect. 3. The performance evaluation is presented in Sect. 4 while the paper is concluded in Sect. 5.

2 System Model

In this section, we firstly give a brief illustration of the AOA-assisted RA. As show in Fig. 1, the cell is divided into several zones based on AOAs. The BS serves each zone via the directional beam, e.g., broadcasting the system information blocks (SIBs) and receiving preambles from MTDs. The angular width (i.e., beamwidth) of each zone is the same. However, once one MTD has multipath spread in neighbor zones, its preamble signals will also be detected by these zones. The MTDs selecting the same preamble in neighbor zones will suffer inter-beam preamble collisions and fail in RA procedures. To remove the influence of multipath angular spread, it is proposed in [9] that the beamwidth of each zone is set to be larger than the maximum multipath spread θ_s as shown in Fig. 2, and the available preamble set is equally divided into two orthogonal subsets (i.e., $\mathcal{P}_1$, $\mathcal{P}_2$ with $\mathcal{P}_1 \cap \mathcal{P}_2 = \emptyset$) that are reused every two adjacent zones. As a result, the MTDs in each zone will not select the same preamble with MTDs in neighbor zones, which can fully avoid the inter-beam preamble collisions. However, the aforementioned alternative preamble reuse scheme in Fig. 1 reduces the number of available preambles for each MTD by half, and thus the preamble collision becomes worse when the access MTDs are concentrated in one zone. Meanwhile, the preambles allocated into neighbor zones with few MTDs are not fully used, which decreases the preamble resource efficiency. Therefore, the fixed preamble allocation is not efficient with various MTD numbers in different zones, and the access performance can be highly improved via dynamic preamble allocation.

The traffic of mMTC can be modeled as the limited Beta distribution proposed by 3GPP [10]. Specifically, N MTDs are assumed to activate in a limited period $[0, T]$. In k-th RA slot as $[t_k, t_{k+1})$, the newly wake-up MTDs is expressed as $M_k =$

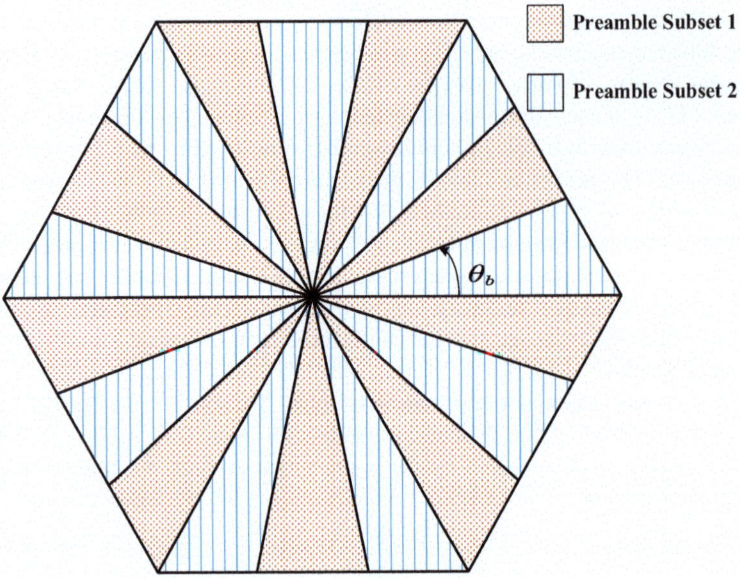

Fig. 1 AOA-assisted RA with spatial preamble reuse, where θ_b represents the beamwidth of each beam

Fig. 2 An illustration of multipath spread

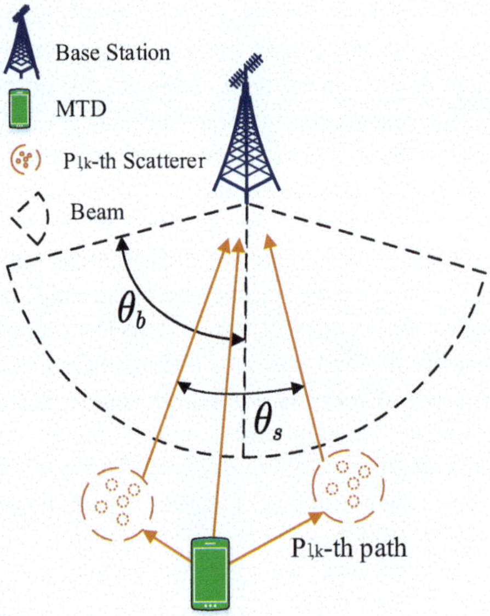

$N \cdot \int_{t_k}^{t_{k+1}} f(t)dt$, where $f(t)$ is the RA attempt intensity as:

$$f(t) = \frac{t^{\alpha-1}(T-t)^{\beta-1}}{T^{\alpha+\beta-1}\text{Beta}(\alpha,\beta)}, \quad \alpha > 0, \beta > 0 \tag{1}$$

where Beta(α, β) represents the Beta function. In this paper, we set $\alpha = 3$ and $\beta = 4$ as [5, 10, 11].

3 Proposed Spatial Dynamic Preamble Soft Allocation

In this section, we propose a novel spatial dynamic preamble soft allocation scheme for AOA-assisted RA to improve access performance and fairness among MTDs.

3.1 Preamble Soft Allocation Framework

As shown in Fig. 1, since non-neighboring zones are allocated with the same preamble subsets, adjusting the preamble numbers of one zone will not only affect its access performance but also other non-neighboring zones'. To explain clearly, we give an example as follow. As shown in Fig. 3, a cell is divided into 6 zones, numbered as Zone 1–6. The preamble subset $\mathcal{P}_1$ is reused in the Zone 1/3/5, and the preamble subset $\mathcal{P}_2$ is reused in the Zone 2/4/6. The MTDs are concentrated in Zone 1 and 4. In this case, with the preamble numbers of $\mathcal{P}_1$ increased, Zone 1 can acquire more available preambles and achieve higher success RA probability. However, with less preambles allocated to $\mathcal{P}_2$, Zone 4 will suffer more severe preamble collision although Zones is not neighbor to Zone 1, which is unfair to Zone 4. Similarly, the success RA probability of Zone 1 decreases with more preambles in $\mathcal{P}_2$ and less in $\mathcal{P}_1$. To improve success RA probability with fairness, we propose that the subset of one zone can be statistically shared to its neighbor zone. The preambles are allocated in statistic among different zones, called soft allocation. The sharing factor is dynamically updated based on the access MTD numbers of the zone and its neighbors, which can be broadcast via SIBs per beam in each RA slot.

We give a further illustration of preamble soft allocation via two typical cases. As shown in Figs. 4 and 5, Zone 1 and 3 are basically allocated with $\mathcal{P}_1$, and Zone 2 with $\mathcal{P}_2$. The beamwidth of each zone is set to be greater than the maximum multipath spread θ_s, and thus the AOA of each MTD can mostly spread in two adjacent beams as shown in Fig. 2. MTDs can identify which zones their AOA spread in based on the receiving power of the corresponding SIBs. In Fig. 4, MTDs are concentrated in Zone 2, and the MTD numbers of Zone 1/3 are small. In this case, the MTDs in Zone 2 can select preambles of $\mathcal{P}_1$ with a given probability (i.e., the sharing factor). In Fig. 5, MTDs are concentrated in both Zone 2 and 4. In this case, $\mathcal{P}_1$ is only statistically

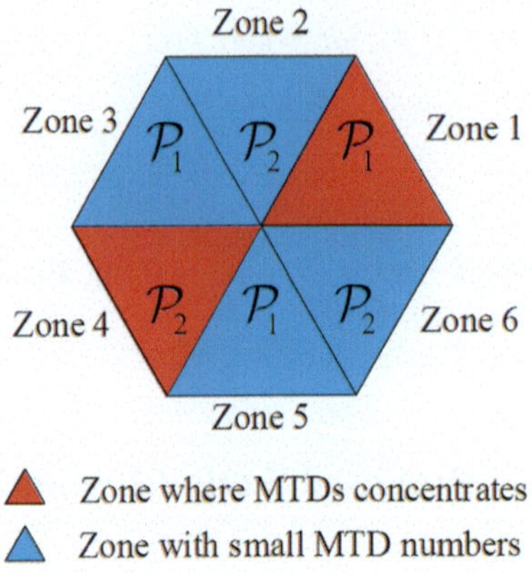

Fig. 3 An example of preamble allocation with orthogonal subsets $\mathcal{P}_1$, $\mathcal{P}_2$

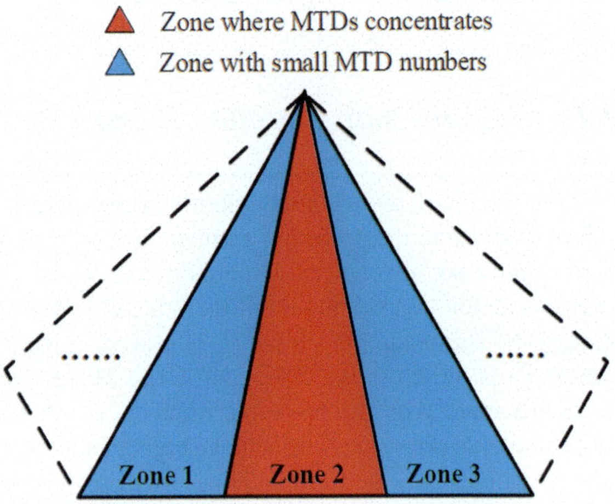

Fig. 4 Case 1

shared to Zone 2's MTDs with AOA spread close to Zone 1. These MTDs can be notified via the SIB of Zone 1, and will not affect Zone 3, which mitigates the effect of multipath spread. Other MTDs in Zone 2 are not allowed to share $\mathcal{P}_1$.

3.2 Dynamic Preamble Soft Allocation Scheme

With a large antenna number in massive MIMO, the front-to-back ratio (FBR) as the ratio of main-lobe to side-lobe gain of each beam can be sufficiently high due to the large array gain. In this case, the receiving power of MTDs in side-lobe is extremely suppressed, and only MTDs that has AOA spread in the main-lobe obtain high receiving power and can be detected via the corresponding beam [9]. The beamwidth of each zone is set to be greater than the maximum multipath spread

Fig. 5 Case 2

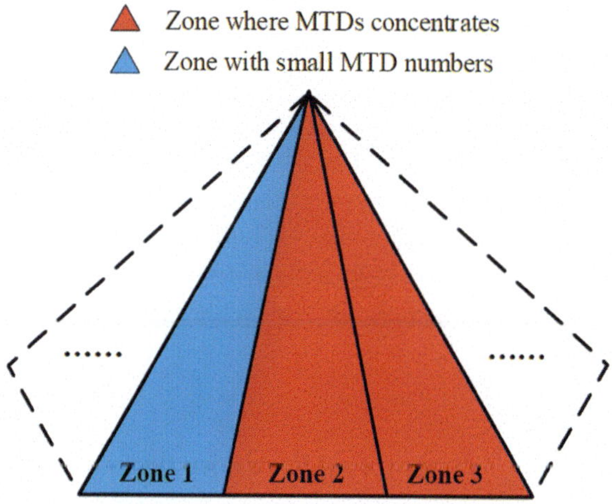

θ_s. The preamble soft allocation information is broadcast to MTDs via SIB per beam. In the proposed scheme, the preamble is also basically divided into two equal subsets. The preamble soft allocation information includes the corresponding basic preamble subset and the statistic shared factors. Since a part of preambles are shared among adjacent zones, the effect of multipath spread is not fully removed. Thus, the BS can not identify the zone that MTDs belong to when the BS detects their preambles. To overcome this shortcoming, we propose to add the zone index to Message 3 (MSG3) of each MTD.

The detailed procedures of the proposed scheme are described as follows:

(Step 0) Configuration Information Broadcast: In each RA slot, the zone index and the updated preamble soft allocation information is broadcast by the BS via SIBs per beam.

(Step 1) MSG1 (Preamble) Transmission: After decoding the SIB, each MTD acquires its available preamble set, which contains the corresponding subset, the shared preambles of the other subset and the statistic shared factor. Then, each MTD randomly selects one preamble from the corresponding subset or the other subset with the given probability based on the statistic shared factor. Each MTD transmits the selected preamble to the BS. The preamble detection is performed per beam at the BS.

(Step 2) MSG2 Transmission: Once the preamble is detected by the BS, a MSG2 containing the random access response (RAR) is replied via the corresponding beam. The RAR contains the zone indexes, the preamble indexes, TA information and uplink resource grant.

(Step 3) MSG3 Transmission: After decoding the MSG2, the MTD matches its corresponding RAR messages by verifying zone index and preamble index. Then, it utilizes the granted uplink resources to transmit a MSG3. MTDs selecting the same preamble will match the same corresponding RAR, and send their MSG3s via the same granted uplink resources. If the AOA of these MTDs are spread in the same beams, their MSG3s will interfere with others, and can not be correctly decoded at the BS.

(Step 4) MSG4 Transmission: The BS replies a MSG4 containing contention resolution message after successfully decoding MSG3 via the corresponding beam. With finishing in decoding the MSG4, the MTD successfully access the network. If one MTD fails in its RA procedure, it will retry in next RA slots until the maximum preamble transmission times is reached.

Note that each beam only performs detection of the corresponding allocated preambles instead of the total preamble set, which reduces the BS processing overhead.

3.3 Soft Preamble Allocation Procedures

In this subsection, we further investigate the procedure of soft preamble allocation. We consider a cell divided into I beam zones, and the beamwidth of each zone is greater than the maximum multipath spread θ_s. N_P available preambles are equally divided into two subsets $\mathcal{P}_1$, $\mathcal{P}_2$, each with $N_s = \frac{N_P}{2}$ preambles. In each zone, $a_i \in [0, 1][0, 1]$ of MTDs have AOA spread in the previous adjacent zones (called P-MTDs), and b_i in the next adjacent zones (called N-MTDs), where $a_i + b_i \leq 1$. The BS can know a_i, b_i from the zone index information included in MSG3, and a_i, b_i can be considered as constants. For simplicity, we assume $a_i = b_i = 0.5$ without loss of generality. The shared factors α_i, β_i are updated in each RA slot, where the P-MTDs of Zone i select the preamble from the other subset with probability α_i, and the N-MTDs with probability β_i. For simplicity, notations are defined as follows: n_i represents the backlogged MTD numbers (containing newly-wake up MTDs and retransmission MTDs) of Zone i in a RA slot. $\mathcal{N}_i \in \{\mathcal{P}_1, \mathcal{P}_2\}$ represents the preamble subset of Zone i, and $\mathcal{N}_{-i}$ represents the other subset. Since Zone I is adjacent to Zone 1, we define that Zone I is also called Zone 0, and Zone 1 is also called Zone $I + 1$. Under this definition, the previous/next adjacent zone of any Zone i is numbered as Zone $i - 1/i + 1$. The Zone i is concentrated distributed with MTDs, and it can share the other subset of Zone $i - 1$ and $i + 1$.

Firstly, we derive the trigger condition of soft allocation based on the RA throughput. The expected number of MTDs that select preambles of the basic corresponding subset $\mathcal{N}_i$ in Zone i is expressed as:

$$Q_i = \frac{n_i}{2} \cdot (1 - \alpha_i) + \frac{n_i}{2} \cdot (1 - \beta_i) = n_i \left(1 - \frac{\alpha_i + \beta_i}{2}\right) \tag{2}$$

Each MTD can succeed in RA when it selects one different preamble from other MTDs of Q_i, and its success RA probability can be calculated as:

$$P_{Q,i} = \left(1 - \frac{1}{N_s}\right)^{Q_i - 1}. \tag{3}$$

L_i/M_i represents the expected number of P-MTDs/N-MTDs selecting preambles of the other subset $\mathcal{N}_{-i}$ respectively, expressed as:

$$L_i = \frac{\alpha_i n_i}{2}, \tag{4}$$

$$M_i = \frac{\beta_i n_i}{2}. \tag{5}$$

Similarly, the success RA probability of L_i or M_i MTDs can be calculated as:

$$P_{L,i} = \left(1 - \frac{1}{N_s}\right)^{L_i + n_{i-1} - 1}, \tag{6}$$

$$P_{M,i} = \left(1 - \frac{1}{N_s}\right)^{M_i + n_{i+1} - 1}. \tag{7}$$

From Eqs. (2)–(7), the expected number of successful RA among n_i MTDs in Zone i can be calculated as:

$$T_i = Q_i \cdot P_{Q,i} + L_i \cdot P_{L,i} + M_i \cdot P_{M,i}, \tag{8}$$

where T_i contains three term and is influenced by the MTD numbers of adjacent zones. From the perspective of each zone, the P-MTDs selecting preambles of $\mathcal{N}_{-i}$ can be equivalently considered as the access MTDs of Zone $i-1$, and similarly N-MTDs selecting preambles of $\mathcal{N}_{-i}$ can be equivalently considered as the access MTDs of Zone $i+1$. Thus, the second term of Eq. (8) can be considered as the part of throughput of Zone $i-1$, and the third term can be considered as the part of throughput of Zone $i+1$. For simplicity, we define the equivalent access MTD number E_i. For Zone $i, i-1, i+1$, the equivalent access MTD number can be expressed as: $E_i = Q_i = n_i[1-(\alpha_i+\beta_i)/2]$, $E_{i-1} = n_{i-1} + \frac{\alpha}{2}n_i$ and $E_{i+1} = n_{i+1} + \frac{\beta}{2}n_i$. When the expected access MTD number is $E_j, j \in [i-1, i, i+1]$ with N_s preambles, the expected RA throughput G_j is calculated as:

$$G_j = E_j \cdot P_{E,j} = E_j\left(1 - \frac{1}{N_s}\right)^{E_j - 1}. \tag{9}$$

By substituting E_i, E_{i-1}, E_{i+1} into Eq. (9), the equivalent RA throughput of Zone $i, i-1, i+1$ can be calculated as:

$$G_{i-1} = \left(n_{i-1} + \frac{\alpha}{2}n_i\right)\left(1 - \frac{1}{N_s}\right)^{n_{i-1} + \frac{\alpha}{2}n_i}, \tag{10}$$

$$G_i = n_i\left(1 - \frac{\alpha_i + \beta_i}{2}\right)\left(1 - \frac{1}{N_s}\right)^{n_i\left(1 - \frac{\alpha_i+\beta_i}{2}\right) - 1}, \tag{11}$$

$$G_{i+1} = \left(n_{i+1} + \frac{\beta}{2}n_i\right)\left(1 - \frac{1}{N_s}\right)^{n_{i+1}+\frac{\beta}{2}n_i}. \tag{12}$$

Since $(1-p)^n \approx e^{-np}, p \ll 1$,[1] Eq. (9) can be approximated as:

$$G_j \approx E_j \cdot e^{-\frac{E_j-1}{N_s}} \tag{13}$$

The first-order partial derivative of G_j *w.r.t.* E_j can be calculated as:

$$\frac{\partial G_j}{\partial E_j} = e^{-\frac{E_j-1}{N_s}}\left(1 - \frac{E_j}{N_s}\right), \tag{14}$$

where the first-order partial derivative G_j *w.r.t.* α_i or β_i can be calculated via derivative chain rule. Since $e^{-\frac{E_j-1}{N_s}} > 0$, the sign of Eq. (14) is decided by $\left(1 - \frac{E_j}{N_s}\right)$. When $\left(1 - \frac{E_j}{N_s}\right) > 0$, G_j monotonously increases with E_j. When $\left(1 - \frac{E_j}{N_s}\right) < 0$, G_j monotonously decreases with E_j. The maximum of G_j can be obtained as $E_j = N_s$, which reveals that the expected throughout of one zone can be improved via adjusting the equivalent expected access MTD number close to the preamble number. If $E_j > N_s$, the expected throughput will decrease with more expected access MTD number. The case $E_j > N_s$ is called overload condition in this paper. If one zone is overload, it can trigger the preamble soft allocation.

Via preamble soft allocation, the overload zone can reduce the expected access MTD number and improve the RA throughout. However, the equivalent access MTD number of adjacent zones will be increased due to P-MTDs/N-MTDs from the overload zone, which decreases the success RA probability of MTDs in these adjacent zones. To improve fairness to MTDs in adjacent zones, sharing preambles of the overloaded zones are forbidden, where the request of preamble soft allocation can be rejected by adjacent zones that are overloaded. The available sharing factor can be calculated as $\alpha_i^{\max} = \min(1, \frac{N_s - n_{i-1}}{a_i \cdot n_i})$ and $\beta_i^{\max} = \min(1, \frac{N_s - n_{i+1}}{b_i \cdot n_i})$. If adjacent zones are overloaded, $\alpha_i^{\max}$, $\beta_i^{\max}$ is negative, which illustrates that PSA is forbidden. If both of these two adjacent zones are not overloaded, the equivalent access MTD number of these adjacent zones can be adjusted as close as possible to further improve fairness in preamble soft allocation. The procedures of preamble soft allocation are illustrated in Fig. 6 as follows.

However, the accurate backlogged MTD number on each Zone is usually not known by the BS. According to [12], the backlogged MTD number can be accurately estimated via Bayesian estimation approach based on the number of idle preambles. In this paper, we assume that the BS can accurately estimate the backlogged MTD numbers and leave the analysis of imperfect estimation for future research.

[1] For example, when $p = 1/64$ and $n = 10$, $(1-p)^n = 0.8543$ is very close to $e^{-np} = 0.8553$.

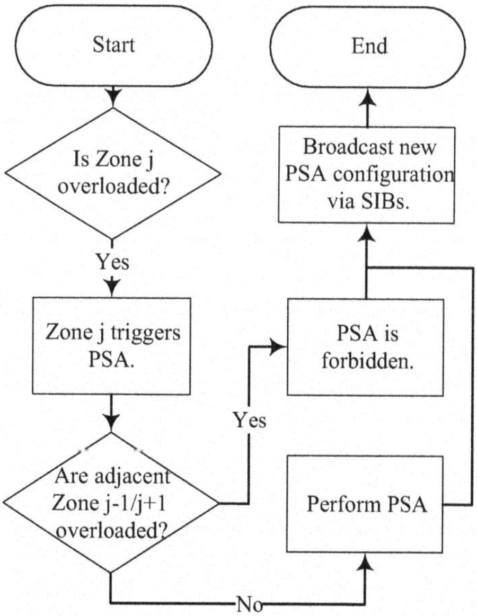

Fig. 6 The procedures for PSA of Zone *j*

4 Performance Evaluation

In this section, we present the performance evaluation of the proposed spatial dynamic preamble soft allocation (PSA) scheme. We consider the beam number is 6, and MTD location are concentrated. The ratio between MTDs in one beam and total MTDs for beam 1 $\sim$ 6 is [5%, 40%, 5%, 10%, 30%, 10%]. The detailed simulation parameters are listed in Table 1. We compare the RA success probability, the average preamble transmission number of successful RA MTDs, and the average preamble transmission number of total RA MTDs of the proposed PSA-RA with the conventional RA scheme and the spatial preamble reuse RA scheme (SFRA) [9]. Specifically, the average preamble transmission times can reflect the average access delay and signaling overhead of RA schemes, which are in proportion to the average number of preamble transmission.

Table 1 Simulation Parameters

Parameters	Values
Total number of MTD with Beta distribution	2000 $\sim$ 1000
Number of preambles	64
PRACH interval time	5 ms
Interval of Beta distribution	10 s (2000 slots)
Number of beams	6
Maximum preamble transmission times	5

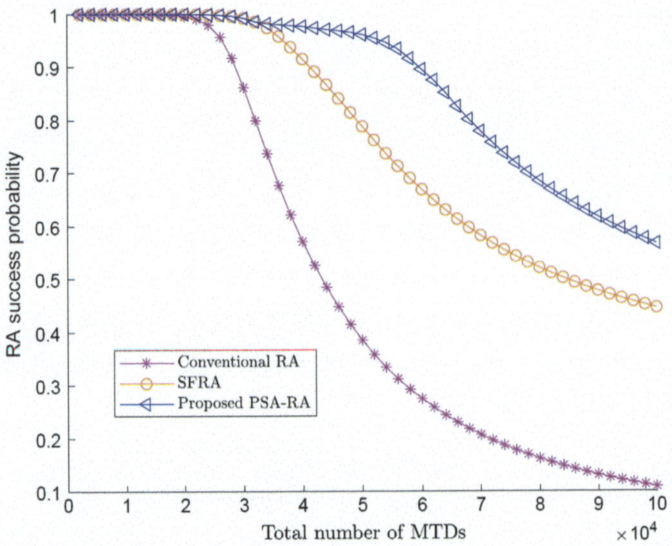

Fig. 7 RA success probability versus the number of MTDs

Figure 7 shows the random access success probabilities of three RA schemes versus the number of MTDs. As illustrated in Fig. 7, the RA success probabilities of three schemes are close to 100% with MTD number, $N < 2 \times 10^4$. With N increasing, the RA success probabilities of these three schemes start to decrease, and the proposed PSA-RA decreases slowly. When $N > 4 \times 10^4$, the RA success probabilities of the proposed PSA-RA is much higher than other two schemes. It shows that the success probability can be improved via the soft preamble allocation.

Figure 8 shows the average preamble transmission times of successful RA versus the number of MTDs. For one successful RA MTD, both the proposed PAS-RA and SFRA need fewer average preamble transmission times, which can reduce the access delay. When $N < 4 \times 10^4$, the success probabilities of the proposed PAS-RA and SFRA are also close to 100% in Fig. 7, and the average preamble transmission times of the proposed PSA-RA is fewer with shorter access delay. The average preamble transmission times of successful RA of SFRA is fewer than the proposed PSA-RA when $N > 5 \times 10^4$. However, in this case, the proposed PSA-RA has much higher RA success probabilities than SFRA, where more than 20% MTDs of SFRA fail in access after retrying preamble transmission at the maximum times.

Figure 9 shows the average preamble transmission times of total MTDs with various MTD number. The proposed PSA-RA has the fewest average preamble transmission times among these three schemes, which reveals that the signaling overhead can also be reduced via PSA-RA. With N increasing, the RA success probabilities degrade, and more MTDs fail in RA. These failed MTDs have transmitted preambles at the maximum times. With the number of failed MTDs increasing, the average

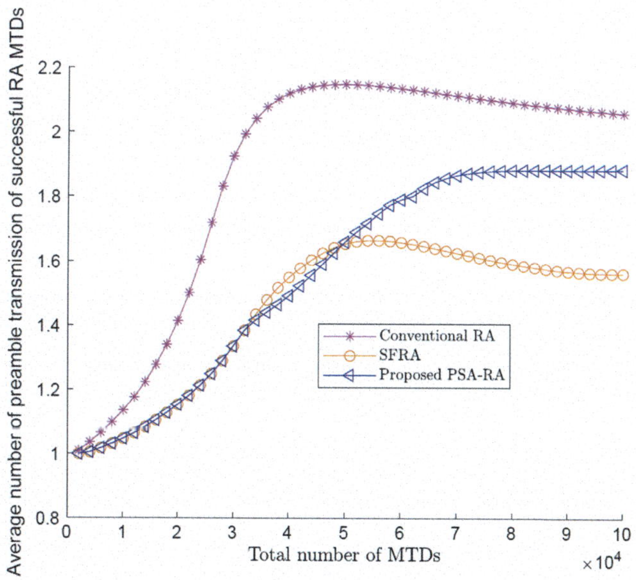

Fig. 8 Average preamble transmission times of successful RA versus the number of MTDs

preamble transmission times of total MTDs of the PSA-RA and SFRA increase and become closer.

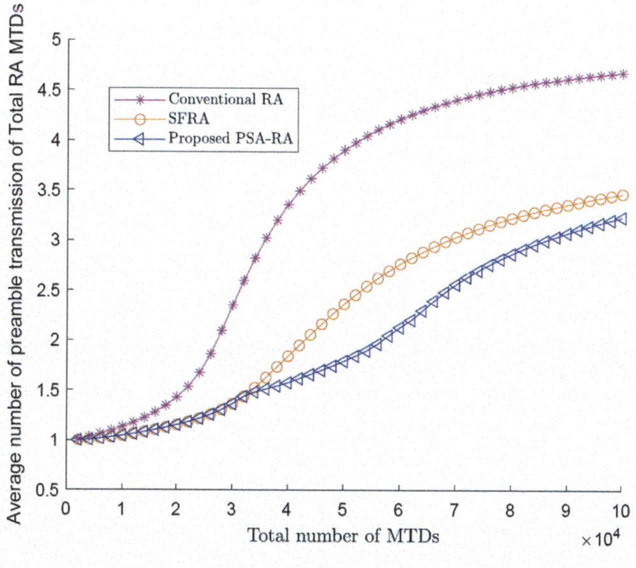

Fig. 9 Average preamble transmission times of Total RA versus the number of MTDs

5 Conclusion

In this paper, we proposed a spatial dynamic preamble soft allocation scheme to improve access performance in the massive MIMO system. In the proposed PSA-RA, preambles can be spatially reused and further statistically shared among MTDs, which improves preamble utilization and user fairness. Simulation results show that the proposed schemes can significantly improve the RA success probabilities and reduce access delay.

Acknowledgements This work was supported by the National Key R&D Program of China under Grant 2018YFB1801103, the Jiangsu Province Basic Research Project under Grant BK20192002, the Key R&D Plan of Jiangsu Province (Prospective and Key Technologies for Industry) under Grant BE2022067 and BE2022067-5, and the Huawei Cooperation Project.

References

1. Chen X, Ng DWK, Yu W et al (2020) Massive access for 5G and beyond. IEEE J Sel Areas Commun 39(3):615–637
2. Dawy Z, Saad W, Ghosh A et al (2016) Toward massive machine type cellular communications. IEEE Wirel Commun 24(1):120–128
3. Cisco U (2020) Cisco annual internet report (2018–2023) white paper. 10(1):1–35. (Cisco, San Jose, CA, USA)
4. 3GPP (2020) Technical specification group radio access network; NR and NG-RAN overall description. 3rd Generation Partnership Project (3GPP),Technical Specification (TS) 38.300, version 16.4.0
5. Duan S, Shah-Mansouri V, Wang Z et al (2016) D-ACB: Adaptive congestion control algorithm for bursty M2M traffic in LTE networks. IEEE Trans Veh Technol 65(12):9847–9861
6. Wang SH, Su HJ, Hsieh HY et al (2013) Random access design for clustered wireless machine to machine networks. In: Proceedings of 2013 First International Black Sea Conference on Communications and Networking (BlackSeaCom). IEEE, Batumi, Georgia, pp 107–111
7. Wang T, Wang Y, Wang C et al (2021) Group-based random access and data transmission scheme for massive MTC networks. IEEE Trans Commun 69(12):8287–8303
8. Kim T, Jang HS, Sung DK (2015) An enhanced random access scheme with spatial group based reusable preamble allocation in cellular M2M networks. IEEE Commun Lett 19(10):1714–1717
9. He Y, Ren G (2020) Spatial filtering based random access for machine-type communications in massive MIMO cellular networks. IEEE Trans Veh Technol 69(10):11479–11490
10. 3GPP (2011) Study on RAN improvements for machine-type communications. 3rd Generation Partnership Project (3GPP), Technical Report (TR) 37.868, version 11.0.0
11. Jeon WS, Seo SB, Jeong DG (2022) POMDP-based contention resolution for framed slotted-ALOHA protocol in machine-type communications. IEEE Internet Things J 9(15):13511–13523
12. Jin H, Toor WT, Jung BC et al (2017) Recursive pseudo-Bayesian access class barring for M2M communications in LTE systems. IEEE Trans Veh Technol 66(9):8595–8599

Seamless Handover in Hybrid VLC and WiFi Network: A Testbed Scenario

Kien Trung Ngo, Fabrizio Giuliano, Stefano Mangione, Tim Farnham, and Ilenia Tinnirello

Abstract In response to the growing demand for wireless communication in high-density environments, Visible Light Communication (VLC) has emerged as a promising complement to Radio Frequency (RF) communication. VLC technology utilizes light-emitting diodes (LEDs) to enable high-speed data transmission, while also providing illumination. In this paper, we propose a hybrid VLC/RF network with multiple VLC access points (APs) under a single WiFi AP. However, frequent handovers are typically required as users move within the network, leading to significant overhead and reducing throughput. To address this issue, we propose an intelligent handover solution based on fuzzy logic (FL) that leverages machine learning algorithms and considers received signal strength indicator (RSSI), channel state information (CSI), and user mobility to enhance handover decisions. We aim to develop a seamless and robust mechanism for horizontal handover (HHO) and vertical handover (VHO) in hybrid VLC/WiFi networks, which we implement in our testbed. Our experiments demonstrate that our proposed approach can achieve maximum handover times at 100 and 400 ms for HHO and VHO, respectively. We have fine-tuned different decision-making models, such as an AdaBoost C4.5 model, which can achieve up to 97.5% accuracy.

K. T. Ngo (✉) · F. Giuliano · S. Mangione · I. Tinnirello
Dipartimento di Ingegneria, Università degli Studi di Palermo, Palermo, Italy
e-mail: kien.ngotrung@unipa.it

F. Giuliano
e-mail: fabrizio.giuliano@unipa.it

S. Mangione
e-mail: stefano.mangione.tlc@unipa.it

I. Tinnirello
e-mail: ilenia.tinnirello@unipa.it

T. Farnham
Toshiba Europe Limited, Bristol, UK
e-mail: tim.farnham@toshiba-bril.com

M. Ma (ed.), *Proceedings of the 12th International Conference on Communications, Circuits, and Systems*, Lecture Notes in Electrical Engineering 1193,
https://doi.org/10.1007/978-981-97-2636-3_8

Keywords Visible light communication · Hybrid architecture · Horizontal handover · Vertical handover · OpenVLC

1 Introduction

The surge in wireless data traffic has recently imposed a significant strain on existing communication networks. As a result, Visible Light Communication (VLC), which utilizes the visible light spectrum for data transmission, has emerged as an alternative communication channel. Meanwhile, WiFi remains the most widely used wireless communication technology for indoor environments. Integrated VLC and WiFi technologies have the potential to create more efficient, reliable, and secure communication systems, leading to benefits for various industries and applications. Hybrid VLC/RF systems are state-of-the-art solutions developed to address mobility/coverage and uplink transmission issues. Typically, the RF link serves as a backup link for VLC coverage gaps or as a VLC feedback and control channel. The advantages of hybrid networks utilizing WiFi and LiFi include high data rates due to the vast spectrum offered by VLC, non-interference with other systems due to the use of a distinct electromagnetic spectrum, and RF's all-encompassing coverage. However, due to users' erratic movements and the short-range nature of VLC equipment, the network would experience continuous handover. Therefore, a intelligent system management is necessary for the hybrid system to optimize all control and operation functions, including seamless handover between VLC and RF networks. The handover solution is what we are focusing on solving.

In this paper, we present a comprehensive study of handover mechanisms in hybrid networks, specifically focusing on the combination of VLC and WiFi networks. Firstly, we provide an up-to-date literature review of current handover work related to hybrid networks. This review provides valuable insights into the existing solutions for seamless handovers in hybrid networks and identifies the gaps that still need to be addressed. Additionally, we propose a robust handover mechanism for hybrid VLC/WiFi networks, which we analyze in our testbed. Our solution considers both horizontal and vertical handovers. Secondly, we verify the effectiveness of our proposal by implementing it in our testbed. We gather user mobility data using the doppler velocity element and utilize both VLC and WiFi networks to make the most accurate handover decision. Finally, we apply fuzzy logic and machine learning algorithms to our testbed, further improving the performance of the handover mechanism.

The paper is organized as follows. In Sect. 2, we provide an up-to-date literature review of the latest handover algorithm for the hybrid system model. Section 3 outlines our proposed handover mechanism, specifically designed to optimize performance in our testbed environment and applicable to other similar hybrid systems. In Sect. 4, we present the results of our experiments and provide a performance evaluation of our proposed handover mechanism.

2 Literature Review

In a hybrid network, horizontal and vertical handovers are the two main types of handovers. Horizontal handover (HHO) occurs within the domain of a single technology, while vertical handover (VHO) occurs between different wireless access technologies. During a VHO, the air interface is modified while keeping the destination's path constant. There are mainly three types of research related to handovers in a hybrid network: (i) HHO in VLC, which focuses on improving performance within the VLC domain; (ii) VHO between VLC and WiFi, which aims to optimize handovers between these two wireless access technologies; and (iii) studies that work on both VHO and HHO, providing a more comprehensive approach to addressing handover challenges in hybrid networks.

Handover metrics are critical in determining when, where, and how to perform handovers in a hybrid network. To ensure seamless user connectivity, various quality of service elements that impact handover must be considered. For example, factors like RSSI, CSI, network load, handover delay/latency, user preferences, and so are all essential considerations. The duration of handovers can also vary depending on the type of handover. However, minimizing handover time is crucial to prevent disruption to user services, and therefore, choosing between HHO and VHO is vital for the hybrid system. Overall, effective handover mechanisms should consider multiple factors, including QoS elements, handover duration, and user preferences. In this paper, we have presented the most recent handover mechanisms for hybrid systems in Table 1. However, it is worth noting that previous studies on HHO or VHO are also relevant to our work and can be found in Ref. [9]. Many of them lack implementation, which is crucial for verifying their effectiveness in a testbed environment. To address this gap, we have developed and implemented our handover mechanism in an open platform. By doing so, we can verify the effectiveness of our proposed handover mechanism in a real-world setting and ensure that it provides superior performance compared to existing solutions.

3 Proposed Method

Handover management is a critical aspect of developing solutions and supporting mobility scenarios. It is the process by which the user maintains its active connection while moving from one point of attachment to another. This section describes the handover process features and provides the handover decision problem in a heterogeneous network. The handover scenarios which we consider are shown in 0 (Fig. 1).

Many works describe the handover process in three phases. Firstly, handover information gathering refers to the process of collecting all the essential information needed to identify when a handover is necessary and to initiate it if required. This

Table 1 Summary of the literature review

References	Outcome	Remark
Vegni and Little [1]	Considered mobility in VLC and HO based on RSI in both overlap and non-overlap coverage	Simulation, HHO
Wu and Hass [2]	A novel handover skipping scheme based on reference signal received power (RSRP) and their changed rate to determine the handover target	Simulation, HHO
Arshad et al. [3]	Proposed HO considers a handover skipping scheme and aims to tackle the negative impact of the handover rate	Simulation, HHO
Purwita et al. [4]	Handover based on RSSI and studied handover probability based on a Markov chain model	Simulation, VHO
Wang et al. [5]	A Markov decision process adopts a dynamic approach to obtain a trade-off between the switching cost and the delay features	Simulation, VHO
Wang et al. [6]	Fuzzy logic (FL) algorithm with dynamic handover scheme for dynamic HO. Channel state information, user speed and desired data rate are considered	Simulation, VHO and HHO
Wang et al. [7]	Adopts a dynamic coefficient via machine learning. Adaptive handover mechanism and selection algorithm optimization are the main focus	Simulation, VHO and HHO
Jarchlo et al. [8]	HO is based on link aggregation and MPTCP tools. The main focus is lower handover outage duration and a high network throughput	Experiments, VHO and HHO

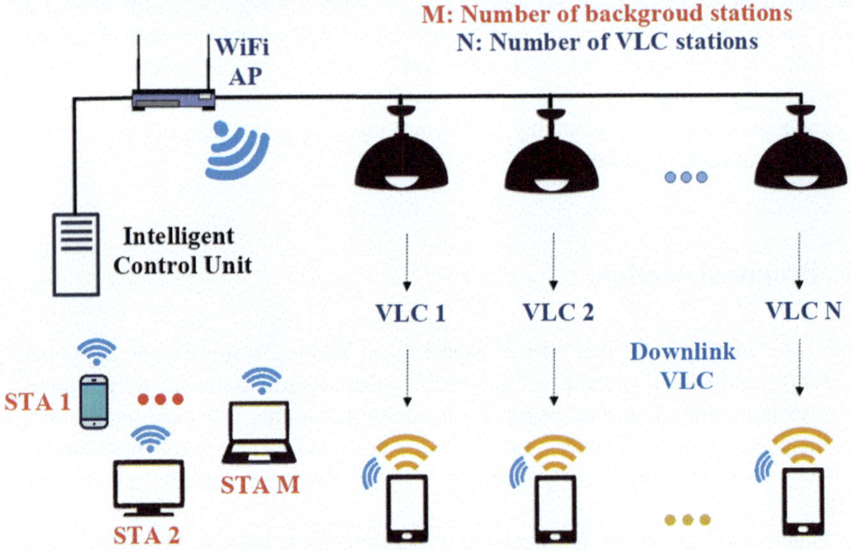

Fig. 1 Simple scenarios for hybrid VLC/WiFI network

phase is also known as the handover initiation or system discovery phase. By gathering critical information, such as signal strength and available bandwidth, this phase helps determine when a handover is necessary and triggers the subsequent steps in the handover process. Secondly, the handover decision phase is responsible for determining whether and how a handover should be performed. This involves selecting the most appropriate access network based on specific criteria, such as user preferences, and providing instructions for the execution phase. Also known as network or system selection, this phase is crucial in ensuring a seamless and efficient handover process. By making informed decisions about which network to connect to next, the handover process can help maximize network performance while minimizing disruptions to the user experience. Thirdly, the handover execution phase involves changing channels in accordance with the details determined during the decision-making phase. Once the appropriate access network has been selected, this phase takes action to execute the handover, ensuring that the user's connection is smoothly transferred from one base station or access router to another. By performing the handover promptly and effectively, this phase helps maintain continuous connectivity for users as they move across different attachment points.

3.1 Handover Information Gathering in VLC

Handover decision criteria help to determine which access network should be chosen, and the handover decision policy represents the influence of the network on when and where the handover occurs. We plan to use RSSI values for traditional handover decision policy because this is the most effective and does not require further information to learn about VLC channel conditions. The technique compares the old RSSI (*oRSSI*) and the new RSSI (*nRSSI*). There are four different policies that can be employed based on these values. (i) RSSI value if *nRSSI* > *oRSSI;* (ii) RSSI with a threshold (*thr)*: choosing the new BS if *nRSSI* > *oRSSI* and *oRSSI* < *thr*; (iii) RSSI with a hysteresis (*h*): choosing the new BS if *nRSSI* > *oRSSI* + *h*; (iv) RSSI with both hysteresis and threshold: choosing the new BS if *nRSSI* > *oRSSI* + *h* and *oRSSI* < *thr* (Fig. 2).

Our testbed is built on OpenVLC 1.3, an open platform that enables researchers to develop their prototypes [10]. The OpenVLC system comprises three components: the OpenVLC cape, which is the front-end transceiver connected to the Beaglebone Black; the OpenVLC firmware, which runs on the Processing Real-time Units (PRUs) that function as microprocessors for the BBB and performs real-time processing; and the OpenVLC driver, which is a module in the Linux kernel that implements the VLC MAC and PHY layers, as well as sampling, symbol identification, coding/decoding, and Internet protocol interoperability. However, the current state of the OpenVLC platform does not provide RSSI to a Linux environment. The OpenVLC firmware stores each of the 16 bits of the current RSSI value in the register of PRU0. Once read, this RSSI is compared with a threshold for decoding. After decoded, a new RSSI value is started to be read, so the previous one is lost. The process shows in 0.

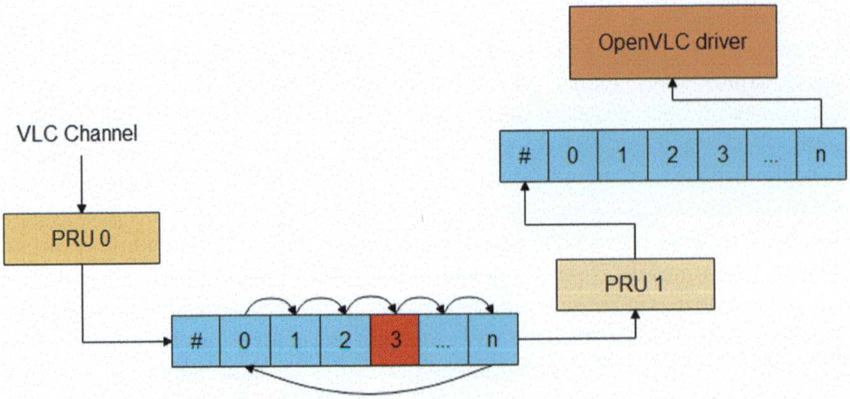

Fig. 2 Memory sharing between PRUs and OpenVLC driver receiver

To extract the RSSI value and integrate it into our handover mechanism, we utilized a prudebug-bbb tool to gain insight into the registers used by the OpenVLC firmware. With this knowledge, we modified the firmware to search for the register containing the RSSI value. Our efforts led us to identify *r3* as the essential register for storing the RSSI value from PRU0. By successfully retrieving the RSSI value on the receiver side, we have taken a significant step toward implementing our handover mechanism.

3.2 Handover Information Gathering in WiFi

We utilize a commercial USB dongle to establish a WiFi channel link in a hybrid system that facilitates testbed replication for fellow researchers. It is crucial to ensure the compatibility of selected dongles with the system and that they meet the necessary performance requirements. Moreover, documenting the configuration process of the dongle and WiFi channel setup could assist other researchers in following the protocol quickly. Our investigation of WiFi channels relied on a diverse set of tools, with particular emphasis on extracting CSI information. In addition to discerning link quality and data rate metrics, we are particularly interested in utilizing CSI information to gain insights into user mobility patterns. As prior studies suggest, doppler velocity information derived from CSI measurements can provide excellent approximations of the user's movement speed [11].

4 Experiments Results

4.1 Hybrid VLC/WiFi Testbed

Our testbed includes two OpenVLC transmitters with WiFi integration, which enable efficient data transmission over visible light shows in Fig. 3. On the receiver side, we have one OpenVLC unit with customized firmware, which is mounted on a robot that can move freely between the two access points. One critical feature of the VLC channel condition is the RSSI value, which measures the signal strength between the transmitter and receiver. To better understand this aspect of the OpenVLC platform, we conducted experiments to observe how the RSSI value changes as the receiver moves further away from the center point of the OpenVLC AP. With all knowledge about OpenVLC, we successfully retrieved the RSSI value on the receiver side, enabling us to integrate it into our handover mechanism, which determines the most effective base station for seamless handover. By conducting experiments to monitor the RSSI value as the receiver moved away from the center point of the OpenVLC AP, we gained insight into the performance of the VLC channel under different conditions. Our results showed that the maximum RSSI value decreased as the receiver moved further away from the center point, indicating a degradation in signal strength like in 0. These observations may also lead to advancements in the performance of VLC systems, as a better understanding of channel conditions can result in more robust systems that deliver higher data rates and improved reliability (Fig. 4).

After careful consideration, we used the sample standard deviation of RSSI values (*sRSSI*) to measure RSSI channel variability. We chose *sRSSI* because it provides a comprehensive view of the RSSI channel and is less prone to bias than other standard deviation estimates. Overestimating variability in samples is preferable to

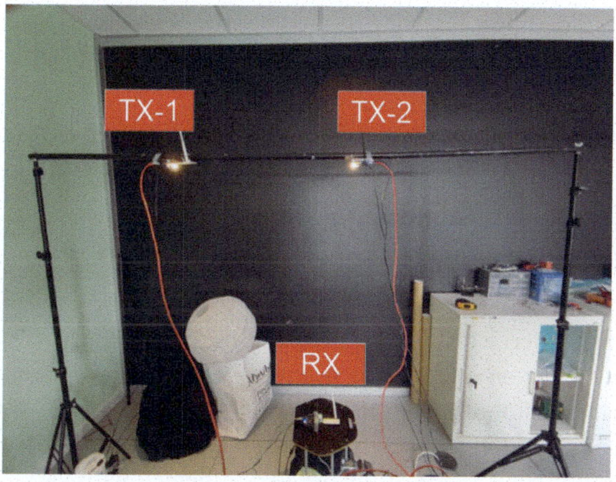

Fig. 3 Experiment setup for hybrid VLC/WiFi testbed

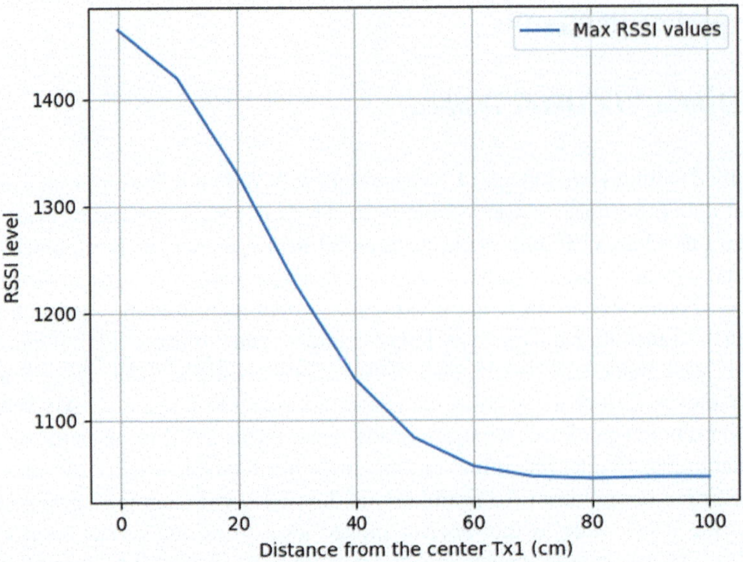

Fig. 4 The maximum RSSI values capture the different distances between center point OpenVLC Tx and OpenVLC Rx

underestimating it, which could potentially lead to inaccurate results. Therefore, *sRSSI* is a reliable way to gauge RSSI variation in our study (Fig. 5).

To investigate user mobility in a WiFi environment, we collected CSI data and extracted doppler velocity components, as depicted in 0. The experimental setup involved linear movements of a receiver at different velocities. The sequence is the receiver moves at high speeds for the first 10 s, followed by a slow movement for the next 10 s, with a 10-s pause in between. This cycle was repeated for 120 s. Our analysis revealed a strong correlation between the doppler velocity and the rate of user mobility, allowing us to categorize it into three levels based on our predetermined thresholds: slow movement (green), medium movement (blue), and fast movement (red). We plan to leverage this information in our machine learning algorithms, which we will elaborate on in the subsequent section in 0.

4.2 *Handover Decision*

Handover decisions are crucial in the handover process, especially when developing a handover mechanism between VLC and WiFi networks. However, this can be challenging due to differences in physical layers, communication protocols, and network architectures. Therefore, selecting appropriate handover criteria is of utmost importance. In our study, we experimented with various combinations of techniques and features that can be collected from both networks. After careful consideration,

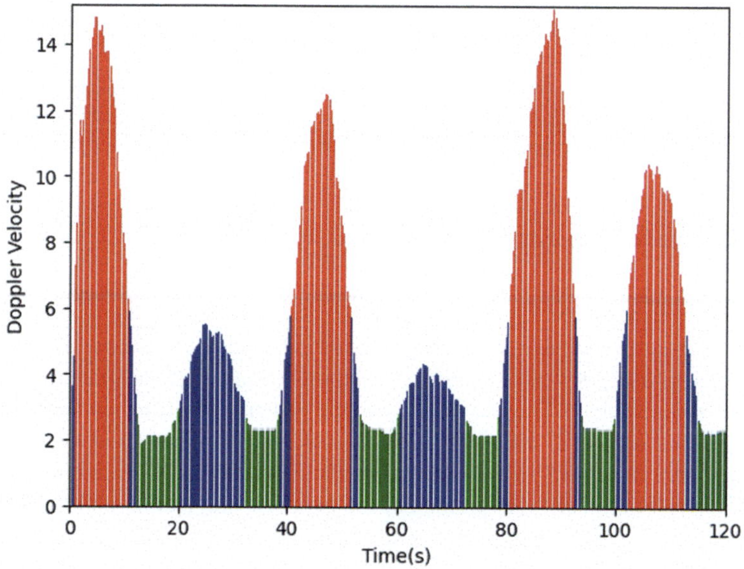

Fig. 5 Doppler velocity envelope extracted from a testbed scenario

we decided to employ fuzzy logic for our small-scale testbed, which allowed us to achieve fast handover times while meeting our requirements.

In our fuzzy logic approach, we consider the *sRSSI 1* and *sRSSI 2* values for two OpenVLC access points, respectively. We choose the WiFi data rate as the most convenient feature to learn about WiFi channel conditions. In our testbed, in case the VLC channel is appropriate for transmitting data, we prioritize it over the WiFi channel, which serves as a backup network. This approach helps reduce the likelihood of bottleneck issues in the WiFi channel [12, 13]. In our OpenVLC platform, the maximum achievable data rate is 400 kbps. We observed that when sRSSI is low, traffic drops below 50 kbps, which led us to define the sRSSI threshold at this level as a handover procedure that should take place. We demonstrate the fuzzy logic rules in Table 2.

Table 2 Fuzzy logic rules

Rule no	Features			AP allocation
	sRSSI 1	*sRSSI 2*	*WiFi data rate*	
1	Not low	–	–	VLC1
2	–	Not low	–	VLC2
3	Low	High	–	VLC2
4	High	Low	–	VLC1
5	Low	Low	Not low	WiFi

Table 3 Handover times

Handover type	Handover time	
	Minimum (s)	*Maximum (s)*
Horizontal	0.043	0.105
Vertical	0.286	0.431

4.3 Handover Execution

As part of our experimental study, we conducted tests in several indoor environments to evaluate the effectiveness of our proposed hybrid handover solution. We collected data for five minutes while the receiver moved in the same cycle as we collected CSI data for doppler velocity. These tests assessed the system's ability to handle handovers seamlessly between VLC and WiFi networks. Our experiments generated valuable data, presented in Table 3, which showcases the minimum and maximum handover times for HHO and VHO. As predicted, the experimental results indicate that the handover time for VHO is typically more than double compared to the handover time in HHO. For HHO, the minimum handover time observed ranges from 0.043 s to 0.105 s, while for VHO, it ranges from 0.286 s to 0.431 s. Although the handover time has not yet been defined in VLC standardization, these results meet the standard for handover time in other radio frequency technologies.

Our results demonstrate better performance than the latest reported handover time for the Li-Wi network in [8] that we discuss in Sect. 2. In that study, the author employed MPTCP and link aggregation tools to develop handover mechanisms, only achieving a minimum handover time of 0.2–0.35 s.

4.4 Handover with User Mobility and Machine Learning

Our experiment aims to scale up our operations in a massive IoT system by utilizing machine learning techniques to develop robust handover systems. Gathering more information about our testbed is crucial before fine-tuning the machine learning model. Moreover, our approach prioritizes decisions based on maintaining a minimum throughput, which guarantees that the quality of service remains consistently high across all user scenarios and movement patterns. To achieve this, we have identified six features–maximum RSSI, minimum RSSI, *sRSSI*, WiFi link quality, WiFi noise level, and WiFi data rate–that can provide valuable insights into network performance. In addition, we have incorporated user mobility as an essential feature by extracting it from doppler velocity data. This information has been categorized into three distinct movement patterns–slow, medium, and fast–as illustrated in 0. By integrating user mobility data with other relevant features, we aim to improve the accuracy and reliability of our machine learning model for predicting handovers in different scenarios (Table 4).

Table 4 Performance of classifiers

Classifier	Training size (%)	Accuracy (%)	Ave. training time (s)
KNN	40	85.4	0.01
	60	90.5	
	80	93.1	
Random forest	40	87.6	0.08
	60	90.3	
	80	95.3	
AdaBoost C4.5	40	87.2	0.03
	60	90.1	
	80	97.5	
LMT	40	86.2	0.05
	60	90.5	
	80	93.1	

To identify the most effective algorithms for improving our machine learning model, we experimented with several supervised learning techniques, including Logistic Regression, Support Vector Machine Classifier, Gradient Boosting, Naive Bayes Classifier, and others. Based on our initial analysis, we selected the four methods that displayed the greatest potential for enhancing the accuracy of our model while requiring a fast-training time. These techniques were K-Nearest Neighbors, Random Forest, AdaBoost C4.5, and Logistic Model Tree. We proceeded to fine-tune these models and evaluated their accuracy results, which are presented in 0.

Our analysis indicates that when 80% of our dataset is used to train the model, the accuracy of predicting handovers exceeds 93%. During the tuning process, we found that the AdaBoost C4.5 algorithm displayed the most significant potential with an impressive 97.5% accuracy. We selected AdaBoost.C4.5 because it employs Decision Trees as its base classifiers, which are widely recognized for their effectiveness in machine learning applications. Specifically, AdaBoost.C4.5 incorporates the C4.5 decision tree algorithm into the AdaBoost framework. Unlike AdaBoost, which uses weak learners, AdaBoost.C4.5 builds a forest of decision trees and selects the best one to add to the ensemble at each iteration. This approach allows AdaBoost.C4.5 to identify more complex relationships in the data than AdaBoost alone, leading to improved predictive accuracy.

Our machine learning model for the handover system was optimized by utilizing two essential hyperparameters, namely *n_estimators* and *learning_rate*. The hyperparameter *n_estimators* was set to 100 and 200, indicating the maximum number of estimators utilized before the AdaBoost.C4.5 algorithm terminated the boosting process. By increasing this value from the default 50, we could reduce the model's variance while preventing overfitting on the training data. Furthermore, we could determine the number of expanding rounds executed before the algorithm stops by utilizing *n_estimators*. The second hyperparameter, *learning_rate*, was tuned from

0.001 to 0.5. This parameter controls the weight update during each iteration of the boosting process. A lower value of *learning_rate* leads to a slower adaptation of the model to the gradient of the loss function, while a higher value results in a faster adaptation. It is crucial to strike a balance between the two hyperparameters since using a small *learning_rate* and many estimators might not lead to better results and can increase computational costs.

In conclusion, selecting appropriate values for these hyperparameters is paramount in influencing the model's performance. As such, we paid keen attention to tuning both hyperparameters to strike a balance between model accuracy and computational efficiency. There is potential to expand this research further by incorporating more transmitter and receiver access points with different hybrid VLC and RF technology. Additionally, we can further boost the performance of our machine learning models by increasing the size of the data available through the hybrid system.

5 Conclusions

Developing a seamless handover mechanism between the two systems is crucial to explore VLC and WiFi integration's full potential. Our proposed testbed serves as a platform for researchers to address the challenges associated with handover mechanisms and develop more efficient, reliable, and secure communication systems across different industries and applications. To achieve seamless integration between VLC and WiFi systems, we proposed a hybrid handover solution based on RSSI and CSI values for both HHO and VHO. Our experimental results indicate that the proposed solution achieves maximum handover times of around 100 and 400 ms for HHO and VHO, respectively. Additionally, our machine learning algorithm for handover decision-making provides remarkable accuracy, with a score of 97.5%. The results demonstrate that our proposed solution can achieve efficient handover times in diverse indoor environments. This underscores the importance of developing a seamless handover mechanism to enable other researchers working on hybrid systems to implement their work in a similar testbed environment.

In the future, several extension ideas warrant further investigation. Integration of tracking and localization into our machine learning model could lead to more efficient handovers by dynamically adapting to user movement patterns and changing network conditions. Developing an architectural solution for managing strip-LED systems allows seamless communication with multiple receivers and traffic flows while balancing the complexity of individual LED intelligence modules. Exploring resource allocation trade-offs between traffic load in each cell and dwell time during handovers may optimize overall system performance. Additionally, investigating adaptive node-association strategies based on anticipated user movement could minimize handover dwell times, even in cell-free systems. By pursuing these research directions, we aim to enhance the applicability and efficiency of our proposed handover mechanism, ultimately contributing to improved user experiences and more robust large-scale networks.

Acknowledgements This work has been funded by the European Union's Horizon 2020 research and innovation programme under the Marie Skłodowska Curie grant agreement ENLIGHT' EM No 814215.

References

1. Vegni AM, Little TDC (2012) Handover in VLC systems with cooperating mobile devices. In: 2012 International conference on computing, networking and communications (ICNC). IEEE
2. Wu X, Haas H (2019) Handover skipping for LiFi. IEEE Access 7:38369–38378
3. Arshad R et al (2016) Handover management in 5G and beyond: a topology aware skipping approach. IEEE Access 4:9073–9081
4. Purwita AA et al (2018) Handover probability of hybrid LiFi/RF-based networks with randomly-oriented devices. In: 2018 IEEE 87th vehicular technology conference (VTC Spring). IEEE
5. Wang F et al (2015) Efficient vertical handover scheme for heterogeneous VLC-RF systems. J Optic Commun Netw 7(12):1172–1180
6. Wang Y, Wu X, Haas H (2016) Fuzzy logic based dynamic handover scheme for indoor Li-Fi and RF hybrid network. In: 2016 IEEE international conference on communications (ICC). IEEE
7. Wang L, Han D, Zhang M, Wang D, Zhang Z (2021) Deep reinforcement learning-based adaptive handover mechanism for VLC in a hybrid 6G network architecture. IEEE Access 9:87241–87250. https://doi.org/10.1109/ACCESS.2021.3089521
8. Jarchlo EA et al (2022) Li-Wi: an upper layer hybrid VLC-WiFi network handover solution. Ad Hoc Netw 124:102705
9. Wu X et al (2021) Hybrid LiFi and WiFi networks: a survey. IEEE Commun Surv Tutor 23(2):1398–1420
10. Galisteo A, Juara D, Giustiniano D (2019) Research in visible light communication systems with OpenVLC1. 3. In: 2019 IEEE 5th world forum on internet of things (WF-IoT). IEEE.
11. Li X et al (2017) IndoTrack: device-free indoor human tracking with commodity WiFi. In: Proceedings of the ACM on interactive, mobile, wearable and ubiquitous technologies, vol 1.3, pp 1–22
12. Ngo KT, Mangione S, Tinnirello I (2021) Exploiting EDCA for feedback channels in hybrid VLC/WiFi architectures. In: 2021 19th Mediterranean communication and computer networking conference (MedComNet). IEEE.
13. Ngo KT et al (2022) Hybrid VLC/WiFi architectures with priority feedback channels. In: 2022 61st FITCE international congress future telecommunications: infrastructure and sustainability (FITCE). IEEE

Design and Implementation of Industrial Internet Emulation Platform Based on Virtualization Technology

Shuai Wang, Wei Li, Wan Bao, Yang Zhao, and Mingxin Lu

Abstract This paper proposes a reconfigurable high-fidelity large-scale industrial Internet emulation platform based on virtualization technology and its implementation method. The emulation platform includes a Logic Control Plane, a Data Plane, and a Data Acquisition and Analysis Plane. The emulation platform generates digital twins of various devices in the Industrial Internet through digital twin technology and build a network through Software-Defined Network (SDN) technology to form a digital twin of the Industrial Internet. The emulation platform can perform offline emulation verification and real-time emulation monitoring of the industrial Internet as required, and display, monitor and analyze the emulation results in real time. Iterative optimization of the digital twin model through feedback of offline and real-time emulation results. The use of SDN technology for networking has the characteristics of flexible network architecture and supports the dynamic addition and rapid deployment of new nodes after network deployment. The emulation platform supports rapid reconstruction of emulation scenarios through industrial big data such as digital twin model library, it also supports the scalability of network emulation nodes by adding servers, it also supports the emulation of large-scale industrial Internet, it also supports the collection of emulation data, evaluates the simulated Industrial Internet according to preset evaluation indicators, and gives optimization suggestions. The emulation platform has the characteristics of high fidelity, scalability and reconfigurability. This paper provides a reliable emulation platform implementation method for the complex, real-time and dynamic industrial Internet environment.

Keywords Industrial Internet · Digital twin · Emulation platform · Reconfigurable · High-fidelity · Large-scale

S. Wang (✉) · W. Li · W. Bao · Y. Zhao · M. Lu
Jiangsu Jinling Science & Technology Group Co., Ltd., Nanjing, China
e-mail: yiruoyun@163.com

M. Ma (ed.), *Proceedings of the 12th International Conference on Communications, Circuits, and Systems*, Lecture Notes in Electrical Engineering 1193,
https://doi.org/10.1007/978-981-97-2636-3_9

1 Introduction

Industrial Internet is a new type of infrastructure, application model and industrial ecology that deeply integrates the new generation of information and communication technology with the industrial industry. Through the comprehensive connection of people, machines, objects, systems, etc., Industrial Internet supports the ubiquitous connection, flexible supply, and efficient allocation of manufacturing resources, and provides a way to realize the development of industrial digitization, networking, and intelligence. The Industrial Internet Emulation Platform is an important platform built for the realization of large-scale equipment interconnection and intelligent manufacturing under the development of Internet technology. It is not only the infrastructure for building the Industrial Internet, but also the core platform that connects people, machines and data.

In the prior art, the industrial Internet emulation platform adopts a predetermined network topology, establishes the connection between hardware through switches and other devices, and then divides the area according to the different functions between the devices. The emulation object is single-solidified and the general network scale is small. It cannot adapt to the networking emulation of large-scale industrial Internet. Moreover, the traditional industrial Internet emulation is an offline emulation, which cannot effectively monitor and simulate the production process in real time. It lacks the description of the characteristics and behaviors of physical entities. There is a big difference between the emulation parameters and the actual industrial parameters. The authenticity and validity of the results cannot be guaranteed. At the same time, due to the single technology of traditional emulation methods and excessive reliance on hardware devices, it is impossible to flexibly reconstruct different emulation scenarios. Nowadays, with the continuous development of the industrial industry, the flexibility, reconfigurability, and scalability of network emulation cannot be satisfied. Moreover, the deployment cost of the emulation platform is high, the cycle is long, and there are problems such as insufficient information sharing and inability to collaborate. In the face of endless new loopholes and attack methods, the firewall, anti-virus and intrusion detection technologies used in a large number of industrial Internet can provide limited security defense. At present, many new solutions have emerged, such as trusted computing, which is related to software and hardware. But the technology lacks the corresponding emulation technology.

This paper provides a multi-level, high-fidelity, scalable, and reconfigurable industrial Internet emulation platform and its implementation method. The emulation platform can be applied to the emulation of large-scale and complex industrial Internet, and the emulation results are real and reliable. In addition, the emulation platform provides a more accurate, comprehensive and low-cost evaluation for the Industrial Internet.

2 Design and Implementation

2.1 Architecture of the Network Emulation Platform

In order to support the emulation of large-scale complex industrial Internet, this paper proposes and designs a reconfigurable high-fidelity large-scale industrial Internet emulation platform and its implementation method based on virtualization technology. It supports offline and real-time emulation and networking through software-defined network technology. The simulated network has the characteristics of reconfigurability and scalability, and at the same time meets the requirements of high-fidelity emulation of the network [1].

In order to achieve the above purpose, the emulation platform designed in this paper includes three parts: Logic Control Plane, Data Plane and Data Acquisition and Analysis Plane, as shown in Fig. 1.

(1) Logical Control Plane: Design a specific industrial Internet scenario consisting of any of the basic components of the industrial Internet such as cloud platform, SDN network equipment, and industrial workshop network according to requirements. Enter the target scene in the Logical Control Plane, and technicians will set the emulation parameters, the parameters are stored in the database, and the main controller issues control instructions according to the target scene parameters to configure and control the simulated network.

(2) Data Plane: simulate and implement the emulation scenarios designed in the Logic Control Plane. The digital twins of industrial production equipment are generated, and a field network is formed through a virtual field bus connection. Several field networks are networked through SDN technology to form a workshop. The workshops and the cloud platform are interconnected through SDN network connections. All these combine to form the digital twin of the Industrial Internet. Among them,

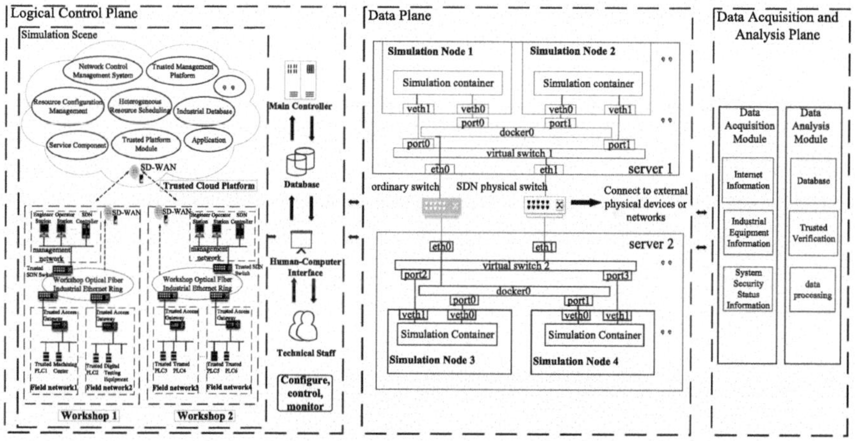

Fig. 1 Schematic diagram of the logic plane of the emulation platform

the virtual switches and real switches are used to form a network for emulation. The virtual switches are connected to virtual devices, and the real switches are connected to the cloud platform and real production equipment to obtain real production data. According to the emulation requirements, the cloud platform can be deployed on the local server to support various applications and management modules required by the industrial Internet cloud platform. In addition, virtual trusted modules are deployed in the digital twins of SDN controllers, SDN switches, PLCs and other devices to simulate trusted startup, dynamic trusted security verification, and trusted device management and control in the Industrial Internet. In the Data Plane, different emulation modes can be selected for offline or real-time emulation according to requirements, and emulation scenarios and emulation data can be displayed in real time [2].

(3) Data Acquisition and Analysis Plane: The plane consists of two parts: a data acquisition module and a data analysis module. The data acquisition module is used to collect offline emulation data and real production data for real-time emulation in the emulation platform, and divide the collected data into network information, Industrial equipment information and system security status information. The collected data is stored in this module. The data analysis module performs data processing and analysis, sets evaluation indicators from two aspects of system performance and system reliability, evaluates the simulated industrial Internet according to the evaluation indicators and gives optimization suggestions, so as to optimize the real and digital twin industrial Internet.

In order to preserve the authenticity of the emulation platform, and at the same time introduce reconfigurability, scalability, and credibility into the emulation platform, use digital twin technology to build emulation scenarios, conduct offline emulation verification and real-time emulation monitoring, the logical behavior function configured by the digital twin and the instructions issued by the main controller are used to coordinately control the industrial Internet digital twin for offline emulation. Real production data is obtained in real time through the real switch, and the data drives the digital twin for real-time emulation [3]. The emulation results iteratively optimize the digital twin model to ensure the high fidelity of the emulation platform. The emulation platform supports the rapid reconstruction of emulation scenarios driven by industrial big data such as digital twin model library and process database, and the software-defined network supports the dynamic addition, rapid deployment and topology planning of virtual nodes, and allows adding virtual nodes by adding servers to realize the rapid reconstruction and expansion of the network.

2.2 Implementation Method

In the implementation process of the reconfigurable high-fidelity large-scale industrial Internet emulation platform based on virtualization technology described in this paper, the following specific configurations are performed:

(1) Main controller configuration: responsible for reading the stored network model parameters, and generating corresponding instructions according to the reading results, scheduling emulation resources in the Data Plane, creating emulation nodes, issuing control instructions, driving various emulation elements, and build the target Industrial Internet.

(2) Software-defined controller and switch configuration: To achieve the separation of control and Data Planes, SDN controllers receive topology transformation information from the main controller, and convert it into a flow table and send it to SDN switches to build a target network. The SDN controller masters the operation of switches and access devices in the entire network, and provides functions such as topology management, switching device management, alarm management, traffic statistics, and mixed traffic transmission control. This emulation platform uses several real SDN switches and virtual SDN switches, and supports the interconnection between real SDN switches and virtual SDN switches. Virtual SDN switches are used to connect virtual devices, and real SDN switches are used to connect external physical devices such as real industrial devices and cloud platforms. Real production data is read through programmable logic controllers (PLC), sensors and other devices into the emulation platform to drive the digital twin for real-time emulation.

(3) Virtual node configuration: Use several servers to implement and configure virtual nodes through the virtual machines in them, and load different digital twins according to requirements. The number of virtualized nodes that can be provided is determined by the hardware performance of the server. The emulation platform supports expanding the network scale by adding servers. Cloud platform functional components, such as industrial database, trusted platform module, resource configuration management module, etc., can be deployed on the local server according to the emulation requirements [4]. The production equipment in the workshop uses the digital twin technology to carry out 3D modeling, lightweight processing, kinematic analysis, establishment of logical functions, establishment of data interfaces, establishment of constraint rules, etc., to generate digital twins corresponding to the equipment, and build networks through SDN technology. The virtual and real SDN network equipment connects the workshop digital twin with the local cloud server to form an industrial Internet digital twin [5]. Virtual nodes can be digital twins of different granularities, such as equipment-level, production-line-level, field network-level, workshop-level, and system-level digital twins.

This emulation platform supports the selection of different emulation modes for offline or real-time emulation according to requirements, including the following four modes: Mode 1 performs offline emulation of the emulation scene. If the target industrial Internet needs to expand the network scale or change the production process, it supports adding new equipment or workshops directly to the emulation scene without affecting the emulation process of other workshops [6]. Mode 2 conducts real-time emulation of the emulation scene, obtains real production data through real switches, drives the industrial Internet digital twin for real-time emulation, and monitors real-time production conditions. Mode 3: In the process of establishing the digital twin of the target industrial Internet, for devices that are difficult to establish a digital twin model, it is supported to connect to real devices through real switches in the emulation platform, and virtual and real devices can be combined for real-time emulation. Mode 4: After the emulation scene is established, perform real-time emulation on the running production workshop, and perform offline emulation first for the digital twin of the workshop that plans to join the Industrial Internet. If the performance and functions meet the expected requirements, a real production scenario is established for manufacturing, and the real production data drives the workshop digital twin for real-time emulation, and the digital twin model is revised by comparing the real data with the offline emulation data. If the expectations are not met, the workshop digital twin model is optimized. The cloud platform can be deployed on the local server through digital twin technology, and the cloud platform can also be connected to the emulation platform through a real switch.

3 Design and Implementation

This chapter takes the shipbuilding industry Internet as the emulation scene, and takes the emulation mode 4 as an example. Combined with the schematic diagram of the logic plane of the emulation platform in Fig. 1 and the overall work flow chart of the emulation platform in mode 4 in Fig. 2, the implementation method and deployment process of the emulation platform are described.

First, design a specific shipbuilding industry Internet emulation scenario, including the network architecture of the shipbuilding industry Internet, the type and quantity of industrial equipment, and the type and quantity of network equipment. In this embodiment, the emulation scene includes three layers of the shipbuilding digital workshop network, the forwarding control layer, and the industrial cloud platform.

Then design the business model in the target shipbuilding industry Internet, including ship production business, ship processing business, data monitoring and management business, data modeling and analysis business, etc., to meet various business needs to the greatest extent, and give full play to the collaboration of the emulation platform, to achieve high-fidelity, reconfigurable, large-scale network emulation requirements. Specifically, the production and processing business may include specific design data, process data, material data, and so on.

Input the network model configuration parameters of the target shipbuilding industrial Internet in the Logical Control Plane and store them in the database. The main controller issues control instructions according to the configuration parameters, schedules the emulation resources of the Data Plane, and builds a emulation scene.

The Data Plane receives the instructions issued by the Logical Control Plane, and simulates the industrial Internet of shipbuilding through digital twin technology.

Among them, for the construction of workshop digital twins, if there is a corresponding digital twin model in the shipbuilding industry database, it can be used directly after parameter configuration; if there is no corresponding model in the shipbuilding industry database, it can be modeled according to the model parameters

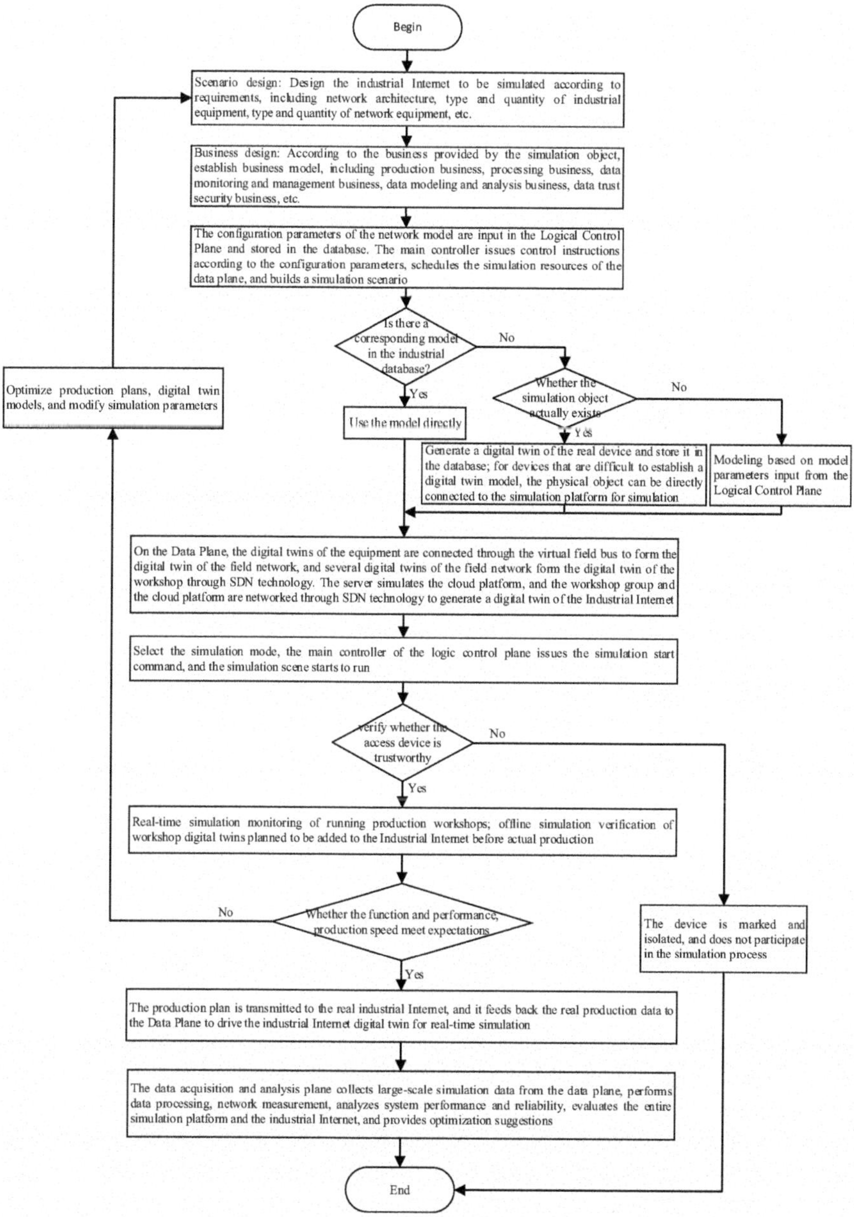

Fig. 2 The overall work flow chart of the emulation platform in mode 4

input by the Logic Control Plane, it can also collect data from people, machines, objects, and environments of the real shipbuilding industry Internet to establish digital models, and generate digital models through lightweight processing, kinematic analysis, the establishment of behavioral logic capabilities, data interfaces, and constraint rules on the generated 3D models. The twin model is stored in the shipbuilding industry database. The equipment digital twin model group adopts a self-organizing cooperation mechanism, and is connected by an industrial field bus to build a field network. Several field networks are networked through SDN technology to form a shipbuilding workshop. Different workshops are interconnected through SDN technology and support interaction with trusted local cloud servers.

Among them, several real SDN switches and virtual SDN switches are used in the Data Plane, and they are connected with each other. Virtual SDN switches are used to connect virtualized nodes, and real SDN switches are used to connect external physical devices, and read real production data through PLCs, sensors, and other devices. The SDN switches transmit the switch data back to the main controller in real time, and the main controller analyzes and displays the network status. The virtual SDN controller receives the topology control instructions to convert the network connectivity status into flow table information in real time, and sends it to the SDN switches. The SDN controller performs unified control and management of the SDN switches, monitors the full flow of the workshop network in real time, and isolates suspicious devices, IP addresses, and ports.

The latency for a flow entry to become effective over the real and/or virtual switches are shown in the Fig. 3. Flow control brings average 25-50ms latency. The emulation performance of link characteristics through real and/or virtual SDN switches, such as link delay, bandwidth and packet loss ratio, are measured in the platform, as listed in Tables 1, 2 and 3. The results show that no matter the characteristics are emulated between two physical nodes through a real SDN switch, one physical node and one docker through a real SDN switch and a virtual SDN switch, and two dockers through a virtual SDN switch, the fidelity of the link characteristics is basically satisfying.

The large-scale on-site network and workshop in the shipbuilding industry Internet are respectively networked through SDN technology, which separates network data and network control, to realize data interaction and collaboration. The SDN network has the characteristics of flexible network architecture, supports the dynamic joining and rapid deployment of new nodes after network deployment, and realizes the reconfiguration of the simulated network. Among them, the cloud platform is deployed on the local server through digital twin technology, and cloud platform components, such as application programs, trusted management platform, network control management, resource management, industrial database, etc., are configured on the local server according to the requirements.

For the equipment that needs to be connected to the shipbuilding industrial Internet, virtual trusted modules are added to the corresponding digital twins, such as trusted management agent modules deployed in trusted SDN controllers, switches, and industrial equipment. Deploy the trusted security management platform in the local cloud server. The physical trust root is simulated by software, and the virtual

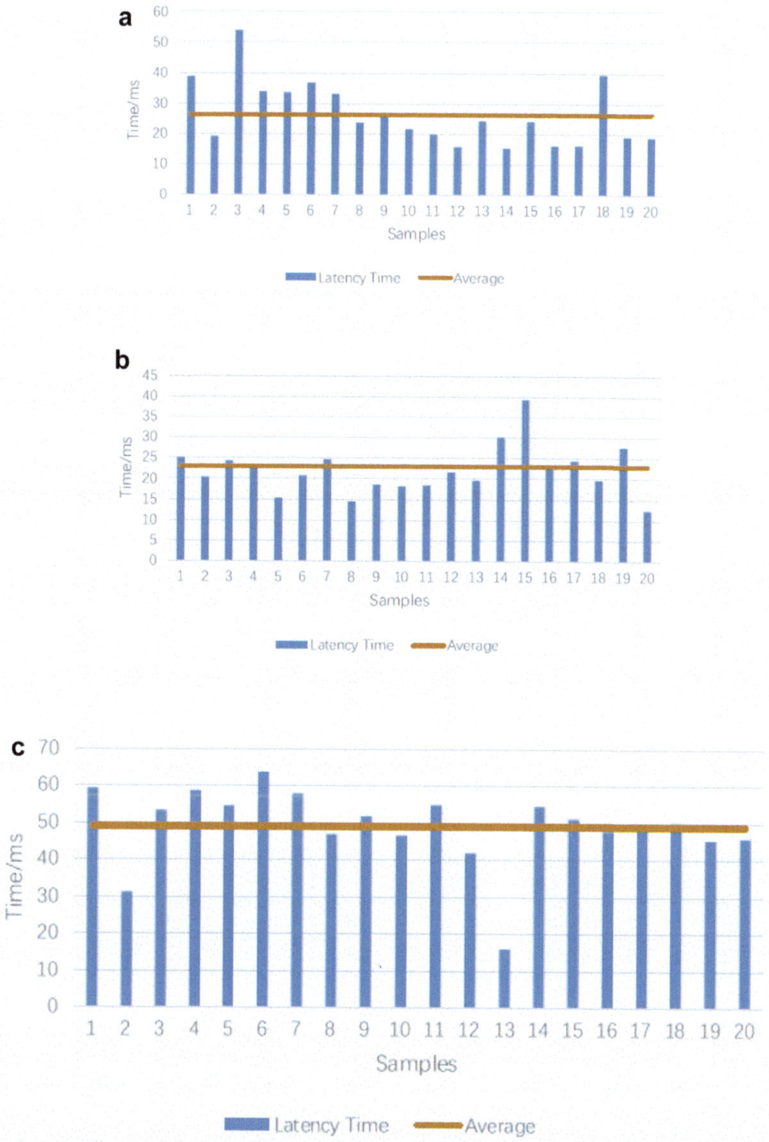

Fig. 3 Latency for a flow entry to become effective on different switches. **a** Latency for a flow entry to become effective on a real SDN switch. **b** Latency for a flow entry to become effective on a virtual SDN switch. **c** Latency for a flow entry to become effective on both a real SDN switch and a virtual SDN switch

Table 1 Performance of link delay emulation between two physical nodes, one physical node and one docker, and two dockers

Link delay set (ms)	Physical node-physical node measured (ms)	Physical node-docker measured (ms)	Docker-docker measured (ms)
5.000	5.079	5.037	5.082
50.000	50.048	50.049	50.109
100.000	100.048	100.046	100.105
500.000	500.051	500.043	500.123
1000.000	1000.050	1000.085	1000.118

Table 2 Performance of bandwidth emulation between two physical nodes, one physical node and one docker, and two dockers

Bandwidth set (Mbps)	Physical node-physical node measured (Mbps)	Physical node-docker measured (Mbps)	Docker-docker measured (Mbps)
0.050	0.046	0.048	0.048
0.500	0.478	0.478	0.477
1.000	0.956	0.956	0.955
500.000	471.038	470.980	471.119
1000.000	938.249	937.330	952.900

Table 3 Performance of link packet loss emulation between two physical nodes, one physical node and one docker, and two dockers

Link packet loss ratio set (Mbps) (%)	Physical node-physical node measured (Mbps) (%)	Physical node-docker measured (Mbps) (%)	Docker-docker measured (Mbps) (%)
1.00	1.00	0.90	0.91
5.00	5.15	4.88	4.53
10.00	10.85	9.73	9.50

trust root is associated with the physical trust root, so that the trust chain is extended from the physical environment to the virtual environment (such as virtual machine), so as to provide the same security function as hardware root of trust for virtual environment, to realize the emulation of the trusted security architecture in the shipbuilding industry Internet.

After the construction of the Internet emulation scene of the shipbuilding industry is completed, the main controller of the Logic Control Plane issues control instructions to the terminal programs of each node in the Data Plane, and the emulation scene starts to run.

The Data Acquisition and Analysis Plane collects large-scale emulation data from the Data Plane, and compares and analyzes the offline emulation data and real-time emulation data in the emulation platform, the real-time emulation data is actually the real production data. The plane then performs further data processing, network measurement, and performance analysis. According to the obtained network information, industrial equipment information and system security status information, the evaluation indicators of system performance and system reliability are scored, and experts are invited to use the hierarchical analysis method to calculate the weight of the indicators of each layer. Combining the index weights with the above evaluation index scores, the evaluation results of the industrial Internet are obtained, and optimization suggestions for the emulation platform and the shipbuilding industrial Internet are provided.

Enterprises can visually view the emulation of the entire shipbuilding industry Internet through the interactive interface, including the three-dimensional scene, emulation data, production progress, evaluation results of the shipbuilding industry Internet.

4 Conclusion

This paper provides a reconfigurable high-fidelity large-scale industrial Internet emulation platform based on virtualization technology and its implementation method. The emulation platform uses digital twin technology for emulation, supports offline emulation verification and real-time emulation monitoring, iteratively optimizes the digital twin model according to the emulation results, and ensures the authenticity of the emulation to a large extent. The emulation platform supports the emulation of large-scale industrial Internet, and supports adding virtual nodes by adding servers. The SDN networking method can realize the dynamic addition of new nodes, rapid deployment and rapid network reconstruction, and can quickly reconstruct the emulation scene according to the digital twin model library and process database. The emulation platform supports the emulation of the trusted computing architecture in the Industrial Internet, deploys trusted modules in the digital twin, realizes the trusted startup and trusted verification of the Industrial Internet, and achieves the protection effect of active immunity. By obtaining network information, industrial equipment information and system security status information as the basis for evaluation, the Industrial Internet will be comprehensively and multi-level evaluated from the two aspects of system performance and system reliability, and optimization suggestions will be provided.

References

1. Zhang Z, Dershan R, Enayati AMS, Yaghoubi M, Richert D, Najjaran H (2022) A high-fidelity emulation platform for industrial manufacturing by incorporating robotic dynamics into an industrial emulation tool. IEEE Robot Autom Lett 7(4):9123–9128
2. Guo Z, Meng W, Wang X, Pan Y, Li R (2020) Models and modeling techniques in the industrial internet. In: Proceedings of the 2020 international conference on computer communication and network security (CCNS), pp 43–46
3. Wang Q, Ma G, He S, Huang Q, Gu W (2020) Industrial intelligent optimization system based on industrial internet platform. In: Proceedings of the 2020 IEEE 11th international conference on software engineering and service science (ICSESS), pp 523–526
4. Delfino LR, Garcia AS, de Moura RL (2019) Industrial internet of things: digital twins. In: Proceedings of the 2019 SBMO/IEEE MTT-S international microwave and optoelectronics conference (IMOC), pp 1–3
5. Suhail S, Hussain R, Jurdak R, Hong CS (2022) Trustworthy digital twins in the industrial internet of things with blockchain. IEEE Internet Comput 26(3):58–67
6. Hou D, Zhao K, Li W (2020) An extensible emulation and network performance analysis system for space information network. In: Proceedings of the 2020 IEEE globecom workshops (GC Wkshps), pp 1–6

RISNet: A DL-Based High Accuracy CSI Feedback Approach for RIS-Aided FDD Systems

Xinyi Tang, Limin Xiao, Zhiqiang, Ming Zhao, Yan Zhang, and Yunzhou Li

Abstract Reconfigurable Intelligent Surfaces (RIS) have become a prominent research topic in recent years. However, downlink channel state information (CSI) feedback in RIS-aided frequency division duplex (FDD) system is a considerable challenge due to the expansion of the cascaded channel dimensions. In this letter, a novel deep learning-based CSI feedback architecture named RISNet is proposed to address this issue. RISNet is designed to operate in situations where the direct link between the base station (BS) and the user equipment (UE) is obstructed, and RIS is employed to direct signals towards the UE. To achieve this, an encoder is placed at the UE end to compress the BS-RIS-UE cascaded channel to a codeword and feed it back to the BS. Before this, the variation in the hybrid domain and principal component mark are designed to reduce the complexity of encoder. At the BS end, a designed decoder is used to recover rich features with various receptive fields from the feedback codeword to reconstruct high-precision CSI. Simulations show that the proposed RISNet exhibits superior accuracy compared to existing work with 3.73 dB lower mean squared error and 0.34 higher cosine similarity, while also displaying less computational complexity on the UE side.

Keywords RIS · Deep learning · FDD · Channel feedback

X. Tang · L. Xiao · Zhiqiang · M. Zhao · Y. Zhang · Y. Li (✉)
Beijing National Research Center for Information Science and Technology, Beijing, China
e-mail: liyunzhou@tsinghua.edu.cn

X. Tang
e-mail: tangxy20@mails.tsinghua.edu.cn

X. Tang · Zhiqiang
Department of Electronic Engineering, Tsinghua University, Beijing, China

M. Ma (ed.), *Proceedings of the 12th International Conference on Communications, Circuits, and Systems*, Lecture Notes in Electrical Engineering 1193,
https://doi.org/10.1007/978-981-97-2636-3_10

1 Introduction

The reconfigurable intelligent surface (RIS) holds great potential as a communication technology for the forthcoming 6G systems [1]. RIS can leverage its programmable metamaterial-based components to manipulate the propagation environment and direct signals towards obstructed terminals [2]. In RIS-aided frequency division duplex (FDD) wireless communication systems, optimizing joint beamforming requires downlink channel state information (CSI) at both the RIS and the base station (BS) given the absence of channel reciprocity [3]. Nevertheless, the feedback overhead becomes impractical as the cascaded channel size rises with the amount of BS and RIS arrays. To address this issue, limited CSI feedback structures are crucial for RIS-aided FDD systems.

Limited research has been performed on CSI feedback for RIS-aided systems. Recent work by Shen et al. [4] proposed a dimension-reduced feedback method based on the single-structured sparsity of the BS-RIS-user equipment (UE) cascaded channel. This work is further extended in [5] by incorporating triple-structural sparsity in the RIS cascaded channel, thereby decreasing the feedback overhead for each UE. In [6], the authors address the problem of bit partitioning between channels of RIS-UE and BS-RIS. The potential of deep learning (DL)-based autoencoders in CSI feedback has been demonstrated, inspiring [7] to propose a scheme for beamforming and discrete phase shift design in RIS-aided systems, which provided a refreshing viewpoint for such scenarios.

Unfortunately, this scheme does not permit complete CSI feedback, as [4] and [5] only consider angle information, [6] only considers channel gain, while in [7] the CSI feedback cannot be separated from beamforming. Therefore, in this paper, a DL-based method named RISNet is proposed for complete CSI compression and reconstruction in RIS-aided FDD systems, which exploits the hybrid sparsity of the cascaded channels for preprocessing. Contributions of our work are listed below.

- RISNet is the first proposed DL-based approach for complete CSI compression and recovery in RIS-aided FDD communication.
- RISNet has a lightweight network architecture on the UE side. Attributed to the restricted computing and storage capabilities of the UE, the hybrid sparsity of the cascaded channel is combined with the principal component marking (PCM) strategy to decrease the computational complexity at the UE side when compared to other neural network (NN) encoder.
- Experimental results demonstrate that RISNet outperforms conventional compressive sensing (CS) techniques and NN-based methods with regard to 3.73 dB lower mean squared error (NMSE) and 0.34 higher cosine similarity.

2 System Model

In this letter, a RIS-aided single-UE[1] FDD system with a blocked straight propagating route between the UE and BS is considered. The RIS consists of a uniform planar array (UPA) with $N_R = N_1 \times N_2$ reflective elements and is implemented to assist the BS in serving the UE. The BS contains N_B uniform linear array (ULA) antennas, while the UE has only one receive antenna. The received signal at the UE is represented as

$$y = \boldsymbol{h}^H \Theta \mathbf{G} \boldsymbol{f} x + n, \tag{1}$$

where y denotes the UE side received signal. x is the transmitted signal. $\boldsymbol{f} \in \mathbb{C}^{N_B \times 1}$ represents the beamforming vector at the BS. $\mathbf{G} \in \mathbb{C}^{N_R \times N_B}$ and $\boldsymbol{h} \in \mathbb{C}^{N_R \times 1}$ denote the BS-RIS and RIS-UE channels respectively. $\boldsymbol{\Theta}$ represents the reflecting diagonal matrix of the RIS as $\boldsymbol{\Theta} = \text{diag}(\boldsymbol{\omega}) \in \mathbb{C}^{N_1 \times N_2}$. $n \sim \mathcal{CN}(0, \sigma^2)$ is noise.

The diagonal matrix property, i.e., $\boldsymbol{h}^H \Theta \mathbf{G} = \boldsymbol{\omega}\, \text{diag}\left(\boldsymbol{h}^H\right) \mathbf{G}$ can be utilized to represent the BS-RIS-UE cascaded channel as

$$\mathbf{H} = \text{diag}\left(\boldsymbol{h}^H\right) \mathbf{G}. \tag{2}$$

Considering the widely used Saleh-Valenzuela channel model [8], the downlink channel $\mathbf{G}$ and $\boldsymbol{h}$ can be expressed as

$$\mathbf{G} = \sum_{l_1=1}^{L_1} \alpha_{l_1} \left(\mathbf{b}_{N_1}\left(\psi_{l_1}^{G_r}\right) \otimes \mathbf{b}_{N_2}\left(\theta_{l_1}^{G_r}\right)\right) \mathbf{b}_{N_B}^H \left(\psi_{l_1}^{G_t}\right), \tag{3}$$

$$\boldsymbol{h} = \sum_{l_2=1}^{L_2} \alpha_{l_2} \left(\mathbf{b}_{N_1}\left(\psi_{l_2}\right) \otimes \mathbf{b}_{N_2}\left(\theta_{l_2}\right)\right), \tag{4}$$

where L_1 and L_2 are the numbers of dominant paths for channel matrices $\mathbf{G}$ and $\boldsymbol{h}$. α_{l_1} and α_{l_2} represent the complex gains for the l_1th and l_2th pathways of BS-RIS and RIS-UE channels. $\psi_{l_1}^{G_r}$, $\theta_{l_1}^{G_r}$ and $\psi_{l_1}^{G_t}$ are the angles of arrival (AoAs) and angles of departure (AoDs) for the l_2th BS-RIS channel. ψ_{l_2}, θ_{l_2} denote AoAs of the l_2th RIS-UE path, $\otimes$ means Kronecker product. $\mathbf{b}_N(\cdots)$ represent the ULA steering vector and is defined as $\mathbf{b}_N(\phi) = \frac{1}{\sqrt{N}}\left[1, e^{j2\pi\frac{d}{\lambda}\phi}, \ldots, e^{j2\pi\frac{d}{\lambda}(N-1)\phi}\right]$, where d represents the distance between adjacent array units, and λ represents the carrier frequency. The interval between is generally set to $d = \lambda/2$.

According to (2) the cascaded channel $\mathbf{H} \in \mathbb{C}^{N_R \times N_B}$ can be expressed as

[1] In this letter, a single UE example is considered for the sake of elucidation, however, it should be emphasized that our method can be easily extended to accommodate multiple-user scenarios.

$$\mathbf{H} = \sum_{l_1=1}^{L_1}\sum_{l_2=1}^{L_2} \alpha_{l_1}^{G}\alpha_{l_2}\mathbf{b}_{N_1}\left(\psi_{l_1}^{G_r} - \psi_{l_2}\right) \otimes \mathbf{b}_{N_2}\left(\theta_{l_1}^{G_r} - \theta_{l_2}\right)\mathbf{b}_{N_\mathrm{B}}^{H}\left(\psi_{l_1}^{G_t}\right). \tag{5}$$

3 The Proposed RISNet Framework

In this section, the entire channel feedback framework of the proposed RISNet is introduced, and then the design of each module, i.e., preprocessing, encoder, decoder and postprocessing, is explained individually.

3.1 The Entire Framework of RISNet

Figure 1 depicts the entire limited CSI feedback framework of RISNet. The spatial domain channel matrix $\mathbf{H}$ is transformed into $\mathbf{H}_a$ by preprocessing at UE. Simultaneously, the indicator vector $\boldsymbol{m}$ is fed back, and the encoder compresses the channel matrix $\mathbf{H}_a$ to a codeword $\boldsymbol{v} \in \mathbb{C}^{M\times 1}$. Upon receiving $\boldsymbol{v}$ and $\boldsymbol{m}$, the decoder and post-processing modules are utilized to reconstruct the spatial domain channel matrix $\hat{\mathbf{H}}$.

The compression ratio (CR) of the system is defined as $\mathrm{CR} = \frac{M}{2N_\mathrm{R}N_\mathrm{B}}$, where $2N_\mathrm{R}N_\mathrm{B}$ is the total dimension of the spatial domain channel and M is the dimension of the feedback codeword.

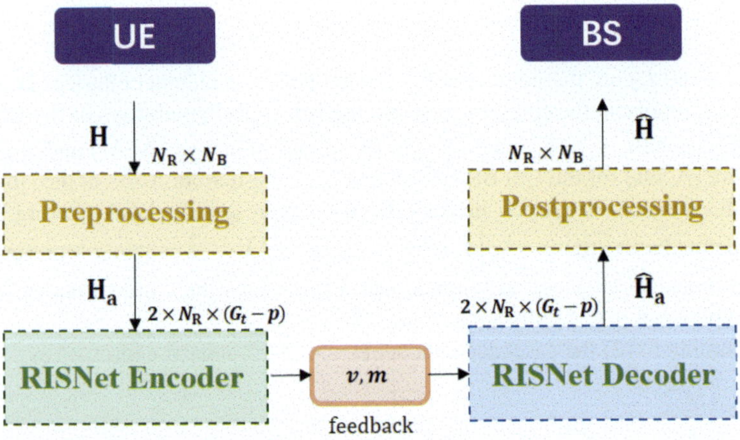

Fig. 1 The workflow of the proposed RISNet

3.2 Preprocessing

Variation in Hybrid Domain In RIS-aided wireless communication systems, BS and RIS are in general located at elevated locations, few obstacles or reflectors exist, resulting in a small number of propagation paths causing scarce scattering, that is, cascade AoAs at the RIS and AoDs at the BS are very limited. As introduced in [4], utilizing a dictionary matrix, the spatial domain channel can be transformed into a hybrid domain channel consisting of both spatiality and angularity. This transformation is formulated as follows:

$$\mathbf{H} = \tilde{\mathbf{H}}\mathbf{F}^H, \tag{6}$$

where $\mathbf{F} \in \mathbb{C}^{N_{\mathrm{B}} \times G_t}$ is the dictionary matrix with each column representing an array response vector corresponding to a certain sampling angle $\bar{\psi}_n$ in space, specifically expressed as $\mathbf{F} = \left[\mathbf{b}\left(\bar{\psi}_1\right), \ldots, \mathbf{b}\left(\bar{\psi}_{N_B}\right)\right]$. When the BS AoD angular resolution G_t equals to N_{B}, $\mathbf{F}$ becomes $N_B \times N_B$ discrete Fourier transformation matrix. Each element of the hybrid domain channel represents the channel component at a particular sampling angle. Therefore, only the angles corresponding to effective scatterers will have non-zero values, with other angles having zero or near-zero values. Considering the L_1 channel pathways between the RIS and BS, only L_1 columns of $\tilde{\mathbf{H}}$ will non-zero, as depicted in Fig. 2. The exact number of columns is affected by the amount of AoDs at the BS [9].

Principal Component Marking The sparse structure of the hybrid CSI channel $\tilde{\mathbf{H}}$, shown in Fig. 2, can be exploited for reducing the size of the channel before feeding it to the encoder. Applying PCM [10] enables a significant decrease in the dimensionality of the input, consequently leading to a lower complexity due to a reduction in the training parameters and computational burden associated with a larger size of input.

The hybrid domain channel can be expressed as $\tilde{\mathbf{H}} = [\boldsymbol{h}_1, \boldsymbol{h}_2 \ldots \boldsymbol{h}_{G_t}]$, where $\boldsymbol{h}_n$ represents the nth column of $\tilde{\mathbf{H}}$. The Euclidean norm of each column is squared and displayed in non-descending order, as represented by $\|\boldsymbol{h}'_1\|_2^2 \leq \|\boldsymbol{h}'_2\|_2^2 \leq \cdots \leq \|\boldsymbol{h}'_{G_t}\|_2^2$.

Then define the residual energy ratio as

$$\gamma_p = E\left\{\frac{\sum_{i=1}^{G_t}\|\boldsymbol{h}'_i\|_2^2 - \sum_{i=1}^{p}\|\boldsymbol{h}'_i\|_2^2}{\sum_{i=1}^{G_t}\|\boldsymbol{h}'_i\|_2^2}\right\}, \tag{7}$$

where $p \in 1, \ldots, G_t$, and ratio γ_p is bounded in the interval [0,1]. Increase p until γ_p less than threshold τ, and keep the remaining $G_t - p$ column signal as a new truncated matrix $\tilde{\mathbf{H}}' \in \mathbb{C}^{N_{\mathrm{R}} \times (G_t - p)}$. A binary indicator vector $\boldsymbol{m} = [10\ldots011]^T \in \mathbb{C}^{G_t \times 1}$ is then generated, where the value of each element indicates whether the related row in $\tilde{\mathbf{H}}$ is retained or eliminated (1 or 0). Afterwards, when the compressed $\tilde{\mathbf{H}}'$ is sent to the decoder for recovery, $\boldsymbol{m}$ is used to ensure that the restored CSI matrix has the same size as the original one.

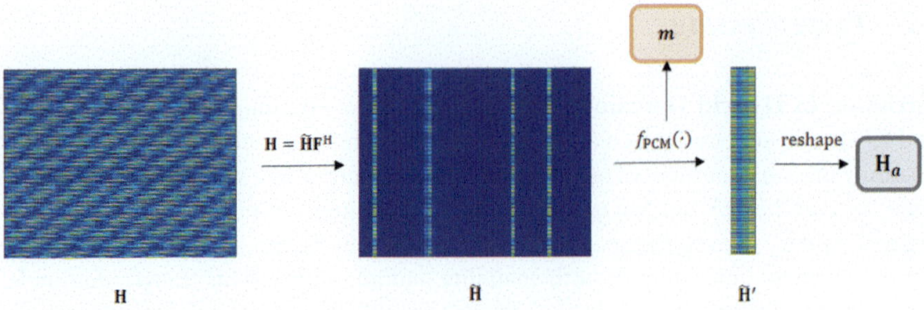

Fig. 2 The preprocessing module

So far we can summarize the preprocessing, as depicted in Fig. 2. The hybrid domain channel $\tilde{\mathbf{H}}$ is transformed from the spatial domain channel $\mathbf{H}$ using the dictionary matrix $\mathbf{F}$. By exploiting the sparsity of the hybrid domain, PCM is employed to selectively discard columns in $\tilde{\mathbf{H}}$ leading to a truncated matrix $\tilde{\mathbf{H}}'$, which reduces the NN complexity. The indicator vector $\boldsymbol{m}$ is sent to the BS. Finally, the truncated channel $\tilde{\mathbf{H}}'$ is reshaped into an image $\mathbf{H}_a \in \mathbb{C}^{2\times N_\mathrm{R}\times(G_t-k)}$ to fit the input requirements of the neural network, with two channels representing the real and imaginary components respectively. The preprocessing is summarized as $(\boldsymbol{m}, \mathbf{H}_a) = \mathrm{f}_{\mathrm{reshape}}\mathrm{f}_{\mathrm{PCM}}(\mathbf{HF}, p))$, which can be denoted as

$$(\boldsymbol{m}, \mathbf{H}_a) = \mathrm{f}_{\mathrm{pre}}(\mathbf{H}, \mathbf{F}, p). \tag{8}$$

3.3 Encoder and Decoder

Encoder To satisfy the limitations of CSI feedback, the encoder is intended to compress the CSI image $\mathbf{H}_a$ into a codeword $\boldsymbol{v}$, as illustrated in Fig. 3. The encoder uses the multi-resolution technique, as presented in [11]. The input undergoes parallel processing. One path consists of three consecutive large-scale convolutional layers designed to provide a wider perception range for the network to effectively handle the sparse CSI matrix. The other path comprises a single 3×3 convolutional layer, which is meant to extract detailed features and manage the dense CSI matrix. Concatenation and 1×1 convolution are used to combined the outputs, followed by a fully connected layer that reduces the features to produce an M-dimensional codeword $\boldsymbol{v}$ based on the given CR.

The encoding process of the system can be expressed as

$$\boldsymbol{v} = \mathrm{f}_{\mathrm{en}}(\mathbf{H}_a, \Omega_{\mathrm{en}}), \tag{9}$$

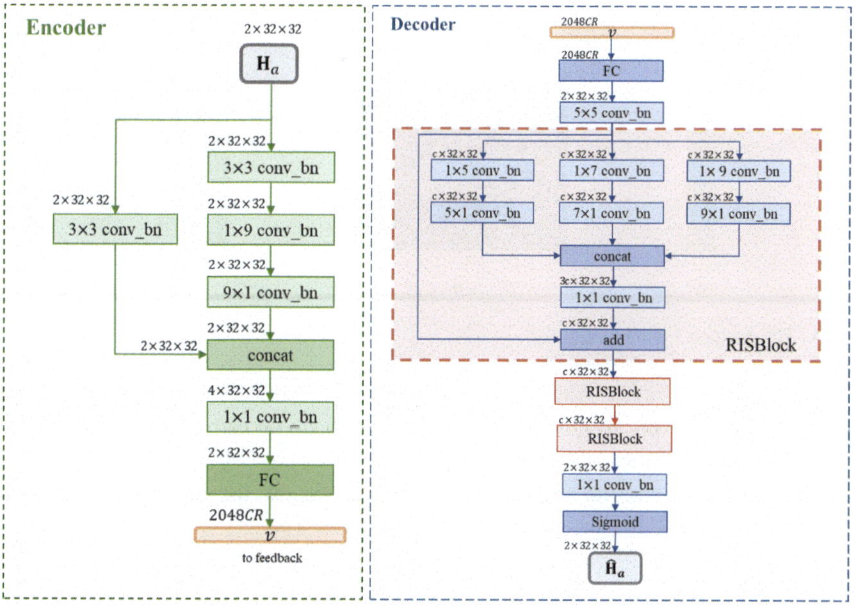

Fig. 3 Design of encoder and decoder in RISNet, where the number of convolution channels, height and width are c, h and w respectively

where f_{en} is the encoder, $\mathbf{H}_a$ is the CSI data to be compressed at the UE side, and Ω_{en} is the parameter set of the NN in the encoder.

Decoder The decoder aims to reconstruct the CSI matrix $\hat{\mathbf{H}}_a$ from the compressed codeword $\boldsymbol{v}$. It starts by resizing the received codeword and extracting rough feature maps through a 5×5 convolution kernel. The central component, the RISBlock, is then deployed and consists of three parallel paths of different resolutions (5, 7 and 9) to extract features with varying receptive fields and reduce computation complexity by decomposing the convolution component. The parallel structure serves to extract additional details from the input. The 'conv_bn' module integrates a convolution with c convolution channels and a batch normalization operation (further information regarding the selection of c and the number of RISBlock n can be found in Sect. 4.2). A 1×1 convolutional layer is employed to the concatenated feature maps along the channel dimension, thereby decreasing the number of convolutional channels from $3c$ to c. This 1×1 convolution is used to increase nonlinearity and fuse features. In reference to ResNet [12], an identity path incorporating the LeakyRule activation function is implemented to counteract gradient disappearance and explosion in the network. After several RISBlocks, another 'conv_bn' reduces the channel size to 2. Finally, the sigmoid is used as the activation function to alter the value range.

The decoding process can be expressed as

$$\hat{\mathbf{H}}_a = f_{de}(\boldsymbol{v}, \Omega_{de}), \tag{10}$$

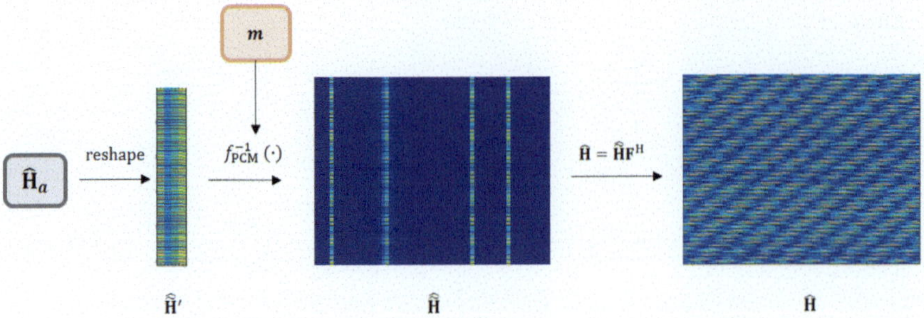

Fig. 4 The postprocessing module

where $\hat{\mathbf{H}}_a$ is the reconstructed CSI matrix, f_{de} is the decoder and Ω_{de} is the parameter of the DL network in the decoder.

The end-to-end learning is applied to dynamically adjust the encoder and decoder parameters in RISNet. This approach can be expressed as

$$\hat{\mathbf{H}}_a = \mathrm{f}_{\mathrm{de}}(\mathrm{f}_{\mathrm{en}}(\mathbf{H}_a, \Omega_{\mathrm{en}}), \Omega_{\mathrm{de}}). \tag{11}$$

As an unsupervised instructional strategy, the parameters Ω_{en} and Ω_{de} of RISNet are optimized according to a meant-square loss function, which can be described as

$$L\left(\Omega_{\mathrm{en}}, \Omega_{\mathrm{de}}\right) = \frac{1}{n}\sum_{i=1}^{n}\left(\mathrm{f}_{\mathrm{de}}\left(\mathrm{f}_{\mathrm{en}}\left(\mathbf{H}_a, \Omega_{\mathrm{en}}\right), \Omega_{\mathrm{de}}\right) - \mathbf{H}_a\right)^2. \tag{12}$$

3.4 Postprocessing

Postprocessing is the reverse of the preprocessing. The recovery matrix $\hat{\mathbf{H}}_a$ is produced by the decoder, and the truncation matrix $\hat{\hat{\mathbf{H}}}' \in \mathbb{C}^{N_{\mathrm{R}}\times(G_t-p)}$ is created after reshaping. Using the binary indicator vector $\boldsymbol{m}$ and the inverse PCM algorithm, the $\hat{\hat{\mathbf{H}}}'$ is converted back to the $\hat{\hat{\mathbf{H}}}$ by adding zeros to the extra rows. Finally, the spatial domain channel $\hat{\mathbf{H}}$ is obtained through the equation $\hat{\mathbf{H}} = \mathrm{f}_{\mathrm{PCM}}^{-1}(\mathrm{f}_{\mathrm{reshape}}(\hat{\mathbf{H}}_a), \boldsymbol{m})\mathbf{F}^H$, which can be simplified as

$$\hat{\mathbf{H}} = \mathrm{f}_{\mathrm{post}}(\hat{\mathbf{H}}_a, \mathbf{F}, \boldsymbol{m}). \tag{13}$$

4 Experiment Results and Analysis

Simulations are performed in this part to illustrate the superiority of our recommended feedback method (Fig. 4).

4.1 Parameter Setting

The total number of UE, RIS elements and BS antennas is set to $K = 1$, $N_\mathrm{R} = 256$ and $N_\mathrm{B} = 64$ respectively, while the BS-RIS and RIS-UE path numbers are defined as $L_1 = 4$ and $L_2 = 4$. All AoAs and AoDs are obtained at random from continuous angular $\mathcal{U}(0, \pi)$ while path gains follow Gaussian distribution $\mathcal{CN}(0, 1)$. For the training process, mean squared error (MSE) as presented in (12) is adopted as the loss function. An NVIDIA 2080Ti GPU was used to train the convolutional layer on Pytorch with a batch size of 256 and 400 epochs, while the Adam optimizer was employed for optimization with an initial learning rate of 0.002 and decay of 75% every 100 epochs. There are 100,000 and 20,000 samples for training and testing sets, respectively. The CR $\in \{\frac{1}{2048}, \frac{1}{1024}, \frac{1}{512}, \frac{1}{256}, \frac{1}{128}, \frac{1}{64}\}$ are used in the simulation. For PCM, the threshold value τ choose 0.67 and 1 two values, the equivalent p values are 60 and 0, respectively.

The accuracy of CSI compression and reconstruction is evaluated using two metrics, the NMSE and the cosine similarity ρ between the initial channel matrix $\mathbf{H}$ and the restored channel matrix $\hat{\mathbf{H}}$ after postprocessing, which can be described as

$$\mathrm{NMSE} = E\left\{\frac{||\mathbf{H}-\hat{\mathbf{H}}||_2^2}{||\mathbf{H}||_2^2}\right\}, \rho = E\left\{\frac{|\hat{\mathbf{H}}^H\mathbf{H}|}{||\hat{\mathbf{H}}||_2||\mathbf{H}||_2}\right\}. \tag{14}$$

The complexity of the neural network is evaluated by floating-point operations (Flops) and the running time required for a CPU to perform channel compression and reconstruction, as mentioned in CsiNet [13].

4.2 Parameter Impact

In this section, the effects of two parameters, the feature map channel size (c) and the number of RISBlocks (n) in RISNet when CR=$\frac{1}{128}$, are analyzed.

It is observed from Tables 1 and 2 that as c and n increase, Flops increases, the network structure becomes wider and deeper, and the characterization ability of the network improves. However, the enhancement of NMSE and ρ shows a marginal decrease. In light of the balance between training consumption and performance, a configuration of 6 convolution channels and 3 RISBlocks is ultimately selected.

Table 1 The impact of channel size c

n	c	Flops (M)	NMSE (dB)	ρ
2	2	1.74	–0.8534	0.4287
	4	2.78	–1.9479	0.6489
	6	**4.37**	**–3.56**	**0.7068**
	8	6.49	–4.0808	0.7759
	10	9.15	–4.2561	0.7904
	32	74.15	–4.6808	0.8359

Table 2 The impact of RISBlock number n

c	n	Flops (M)	NMSE (dB)	ρ
6	1	3.02	–1.7666	0.5829
	2	4.37	–3.56	0.7068
	3	**5.71**	**–4.4427**	**0.7432**
	4	7.06	–4.5504	0.7495
	5	8.4	–4.5903	0.7545
	10	15.13	–4.8747	0.795

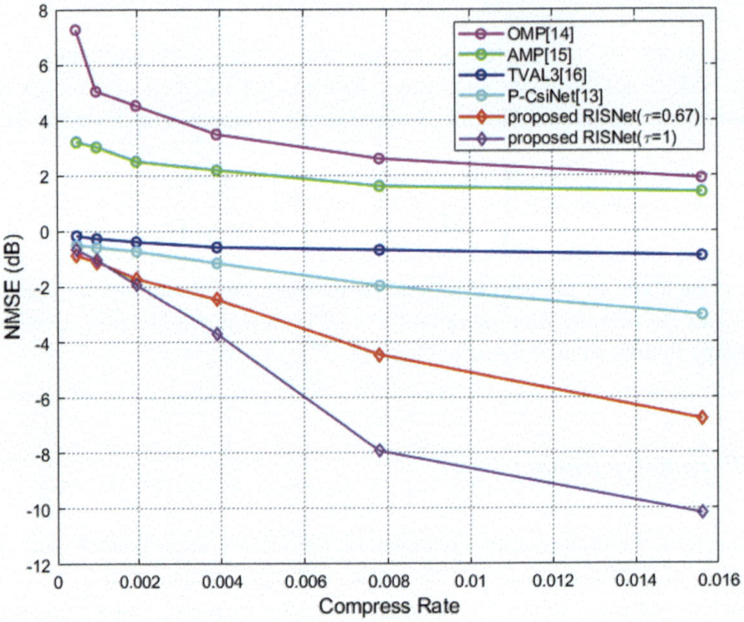

Fig. 5 The NMSE(dB) of various methods

4.3 Performance of the Proposed RISNet

The suggested approach is contrasted to several well-established benchmark algorithms, including the Orthogonal Matching Pursuit (OMP) [14], Approximate Message Passing (AMP) [15], Total Variation-based recovery method TVAL3 [16], and a DL-based method, CsiNet [13]. To ensure a fair comparison, preprocessing and postprocessing (with $\tau = 0.67$) is also performed for CsiNet, which is reported as 'P-CsiNet'.

Based on the results of NMSE and ρ performance demonstrated in Figs. 5 and 6, the proposed RISNet outperforms all other methods across a range of CRs. It was revealed that, when CR is small, i.e. smaller than $\frac{1}{512}$, RISNet shows slight advantages over P-CsiNet [13] and TVAL3 [16], and marked superiority over OMP [14] and AMP [15]. As CR increases, this distinction becomes even more evident, with NMSE and ρ performance for RISNet continuing to improve, while those for AMP [15] and TVAL3 [16] nearly cease to improve with CR larger than $\frac{1}{256}$. Notably, for CR=$\frac{1}{128}$, RISNet showed an impressive NMSE of 3.73 dB lower, and ρ of 0.34 higher than P-CsiNet, demonstrating its profound accuracy in CSI recovery. Additionally, an increase in channel size or the number of RISBlocks, as observed in Sect. 4.2, can further enhance CSI recovery performance.

The effect of preprocessing and postprocessing is evaluated by comparing $\tau = 0.67$ and $\tau = 1$, which denote the dropped columns in the PCM operation for $p = 60$

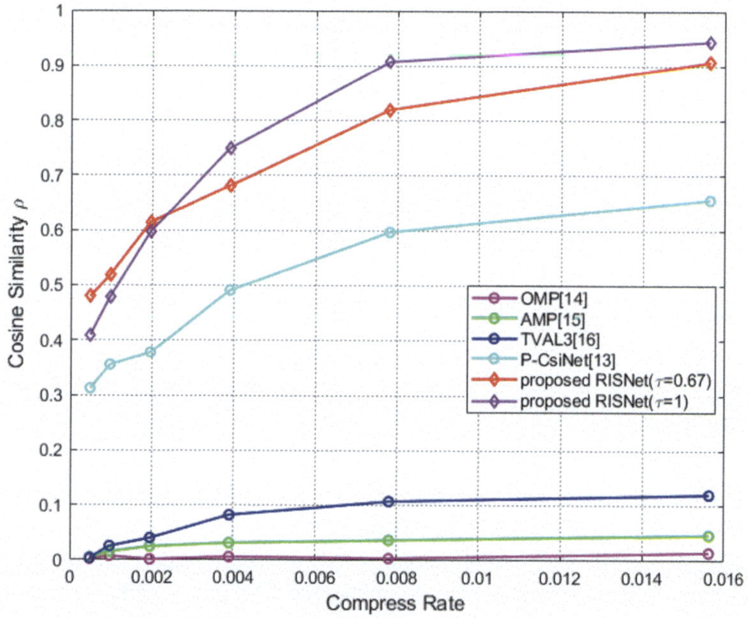

Fig. 6 The cosine similarity ρ of various methods

Table 3 Detailed complexity of different methods

Method	P-CsiNet [13]	RISNet (τ= 0.67)	RISNet (τ = 1)
Params UE	1.05 M	**524 K**	8.39 M
Params BS	1.05 M	**530 K**	8.43 M
Flops UE	1.05 M	**710 K**	11.37 M
Flops BS	**4.54 M**	5 M	80.02 M

and 0, respectively. The smaller the value of p, the fewer the information loss in the channel, but the complexity of the neural network increases. When CR is very low ($\frac{1}{2048}$ and $\frac{1}{1024}$), the performance of RISNet with $\tau = 0.67$ surpasses that of RISNet with $\tau = 1$. Nevertheless, as CR increases, RISNet with $\tau = 1$ starts to demonstrate an advantage in terms of NMSE and ρ performance. Despite this, the complexity of RISNet with $\tau = 1$ is considered too high, making RISNet with $\tau = 0.67$ a more practical option. By adjusting the value of τ, it is possible to balance the trade-off between complexity and CSI compression and recovery performance, with a low value of τ favored if complexity is a concern and vice versa.

4.4 Analysis of Computational Complexity

The average execution time (in milliseconds) required to perform the entire channel compression and reconstruction process on the CPU for several algorithms, including OMP [14], AMP [15], TVAL3 [16], CsiNet [13] and RISNet($\tau = 0.67$), are 212, 126, 384, 3.5 and 3.5, respectively. The results show that DL-based methods, such as RISNet and P-CsiNet [13], are much faster, by about 36–110 times, compared to other CS-based algorithms. Additionally, Table 3 indicates that RISNet($\tau = 0.67$) has significantly reduced the total number of NN parameters required at the UE side, by 49% compared with P-CsiNet [13], thereby exhibiting lower storage capacity requirements. Moreover, the Flops of UE side is 30% lower than P-CsiNet [13], indicating that the proposed method has achieved a lightweight design.

5 Conclusion

In this letter, we have presented an efficient and innovative DL-based RISNet for RIS-aided FDD systems with limited CSI feedback.The proposed architecture comprises an encoder at the UE side, which adopts a hybrid sparsity approach for cascaded channel and PCM preprocessing to compress the CSI into a codeword, and then transmits it to the BS. Additionally, the decoder at the BS has been designed with a multi-resolution path, convolution decomposition, and residual network to recover the CSI from the codeword. Our simulation results indicate that our proposed RISNet outperforms existing methods in terms of compressing and recovering complete CSI. Moreover, our approach offers a lightweight design on the UE side compared to other NN-based methods. In future research, we will investigate the application of quantization enhancement in practical feedback schemes, as well as the integration of feedback schemes with beamforming and discrete phase shift design in RIS-aided systems.

Acknowledgements This work was supported by the National Natural Science Foundation of China under Grant 61631013 and the Key Research and Development Program of China under Grant 2020YFB1806603.

References

1. Yuan X, Zhang Y-JA, Shi Y, Yan W, Liu H (2021) Reconfigurable-intelligent-surface empowered wireless communications: challenges and opportunities. IEEE Wirel Commun 28(2):136–143
2. Tang W, Chen MZ, Chen X, Dai JY, Han Y, Di Renzo M, Zeng Y, Jin S, Cheng Q, Cui TJ (2020) Wireless communications with reconfigurable intelligent surface: path loss modeling and experimental measurement. IEEE Trans Wirel Commun 20(1):421–439
3. Rakhmania AE, Taufik M, Hariyadi A, Hudiono H, Perdana RHY (2021) Per user based interference alignment for uplink comp with perfect csi under different channel models. J Commun 16(10):444–449
4. Shen D, Dai L (2021) Dimension reduced channel feedback for reconfigurable intelligent surface aided wireless communications. IEEE Trans Commun 69(11):7748–7760
5. Shi X, Wang J, Song J (2022) Triple-structured sparsity-based channel feedback for RIS-assisted MU-MIMO system. IEEE Commun Lett 26(5):1141–1145
6. Chen W, Wen C-K, Li X, Jin S (2021) Adaptive bit partitioning for reconfigurable intelligent surface assisted FDD systems with limited feedback. IEEE Trans Wirel Commun 21(4):2488–2505
7. Guo J, Chen W, Wen C-K, Jin S (2022) Deep learning-based two-timescale csi feedback for beamforming design in ris-assisted communications. IEEE Trans Veh Technol
8. Heath RW, Gonzalez-Prelcic N, Rangan S, Roh W, Sayeed AM (2016) An overview of signal processing techniques for millimeter wave MIMO systems. IEEE J Sel Top Signal Process 10(3):436–453
9. Shen D, Dai L (2022) Multi-beam design for extremely large-scale ris aided near-field wireless communications. In: Proceedings of GLOBECOM IEEE global communications conference. IEEE, pp 1–6
10. Chen X, Deng C, Zhou B, Zhang H, Yang G, Ma S (2021) High-accuracy CSI feedback with super-resolution network for massive MIMO systems. IEEE Wireless Commun Lett 11(1):141–145
11. Lu Z, Wang J, Song J (2020) Multi-resolution CSI feedback with deep learning in massive MIMO system. In: Proceedings of International Conference on Communications (ICC). IEEE, pp 1–6
12. He K, Zhang X, Ren S, Sun J (2016) Deep residual learning for image recognition. In: Proceedings of IEEE conference on computer vision pattern recognition (CVPR), pp 770–778
13. Wen C-K, Shih W-T, Jin S (2018) Deep learning for massive MIMO CSI feedback. IEEE Wirel Commun Lett 7(5):748–751

14. Alkhateeb A, El Ayach O, Leus G, Heath RW (2014) Channel estimation and hybrid precoding for millimeter wave cellular systems. IEEE J Sel Top Signal Process 8(5):831–846
15. Bayati M, Montanari A (2011) The dynamics of message passing on dense graphs, with applications to compressed sensing. IEEE Trans Inf Theory 57(2):764–785
16. Li C, Yin W, Zhang Y (2013) Tval3: Tv minimization by augmented Lagrangian and alternating direction algorithm 2009

Electronic and Communication Engineering

Design Step for Op Amp Based CMOS Feedback Amplifiers

Suriya Adirek, Chaiyan Chanapromma, and Jirayuth Mahattanakul

Abstract This paper presents the design step for op amp based CMOS feedback amplifiers. The proposed design step follows the design strategy in which the closed-loop gain and bandwidth of the amplifier are the performance parameters to be satisfied, not op amp's opened-loop phase margin and unity-gain frequency, as traditionally carried out. To demonstrate, the feedback amplifiers were designed for a gain bandwidth (GBW) of 20 MHz and a quality factor (Q) of 0.707. Simulations using 0.35 μm CMOS technology with a 3-V power supply for the feedback amplifiers, obtained from the proposed design steps, show a consumption of significantly less power (45.42%) than those obtained from the traditional design method, while other performance parameters remained comparable.

Keywords Inverting · Feedback · Procedure · Amplifier · Op Amp

1 Introduction

This operational amplifier (op amp) based feedback amplifiers shown in Fig. 1 are widely used in various integrated analog and mixed-signal processing applications. Traditionally, op amps employed in feedback amplifiers are designed to have a phase margin of around 65° [1–9] regardless of the required amplifier gain. However, this

S. Adirek (✉)
Faculty of Agricultural Technology and Industrial Technology, Nakhon Sawan Rajabhat University, Nakhon Sawan, Thailand
e-mail: Suriya.a@nsru.ac.th

C. Chanapromma
Faculty of Industrial Technology, Uttaradit Rajabhat University, Muang Uttaradit, Thailand
e-mail: chaiyanc@uru.ac.th

J. Mahattanakul
The Electrical Engineering Graduate Program, Faculty of Engineering and Technology, Mahanakhorn University of Technology, Bangkok, Thailand
e-mail: jirayut@mut.ac.th

M. Ma (ed.), *Proceedings of the 12th International Conference on Communications, Circuits, and Systems*, Lecture Notes in Electrical Engineering 1193,
https://doi.org/10.1007/978-981-97-2636-3_11

choice would not be an optimal except in the case of an amplifier with a unity feedback factor (100% feedback), such as a unity-gain voltage buffer. For instance, it can be demonstrated that for a maximally flat frequency response (corresponding to the quality factor $Q = 0.707$ as shown Fig. 2, the phase margin of an op amp in feedback amplifiers with a feedback factor of 1/3 should be around 44° [10]. Hence, in such a case, using op amp with a 65° phase margin would be overcompensation and wasteful of power. Consequently, to employ op amps in fixed-gain feedback amplifiers in specific integrated circuits, phase margin, and unity-gain frequency should replace closed-loop gain and bandwidth as performance parameters.

Op amp design steps, first proposed in [11] and later extended in [6, 7, 12–14], enable the designing of an op amp with open-loop gain that satisfies numerous performance parameters, such as phase margin, unity-gain frequency, slew rate, voltage range, noise level, and power consumption. In this paper, the design step compatible with the new op amp design strategy outlined in [10], which emphasizes closed-loop gain, is presented.

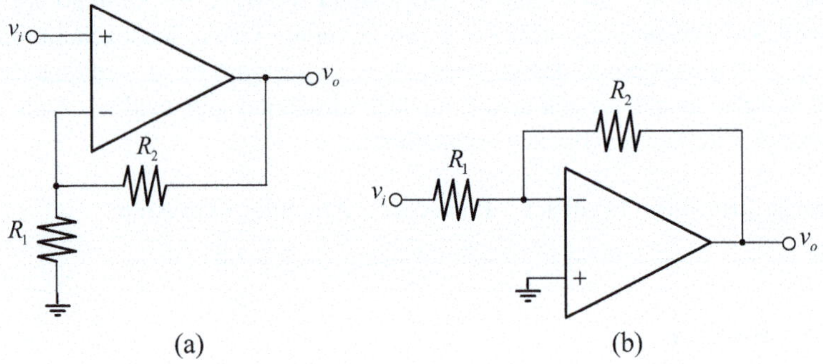

Fig. 1 Op amp based **a** non-inverting amplifier and **b** inverting amplifier

Fig. 2 Frequency responses of a low-pass transfer function with various values of quality factor

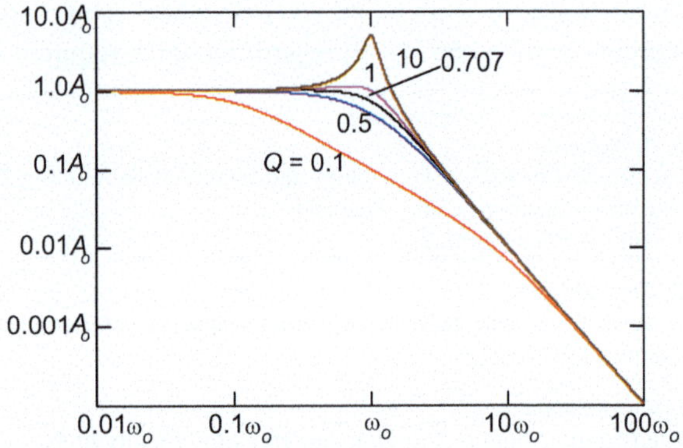

2 Relationship Between Phase Margin and Quality Factor of Feedback Amplifier

The transfer function of an op amp with one dominant pole and one non-dominant pole can be expressed as

$$A(s) = \frac{A_0}{\left(1 + \frac{s}{\omega_{dp}}\right)\left(1 + \frac{s}{\omega_{np}}\right)} \tag{1}$$

where ω_{dp} and ω_{np} represent the dominant pole frequency and the non-dominant pole frequency, respectively. For frequencies higher than ω_{dp}, the Eq. (1) can be re-expressed as

$$A(s) \cong \frac{\omega_{GBW}}{s\left(1 + \frac{s}{\omega_{np}}\right)} \tag{2}$$

where $\omega_{GBW} = A_0\omega_{dp}$, known as the gain-bandwidth product. Neglecting loading effects at the output of op amps, the transfer functions of the op am based non-inverting amplifier in Fig. 1a and the op amp based inverting amplifier in Fig. 1b are given by

$$\frac{v_o(s)}{v_i(s)} = \frac{1}{B}\left(\frac{A(s)}{\frac{1}{B} + A(s)}\right) \tag{3}$$

and

$$\frac{v_o(s)}{v_i(s)} = \left(1 - \frac{1}{B}\right)\left(\frac{A(s)}{\frac{1}{B} + A(s)}\right) \tag{4}$$

respectively. Here, A(s) is the open-loop transfer function of the op amp, and

$$B = \frac{R_1}{R_1 + R_2} \tag{5}$$

represents the feedback factor of the amplifier. Note that the DC gain of the Eqs. (3) and (4) is denoted as

$$\frac{1}{B} = 1 + \frac{R_2}{R_1} \text{ and } 1 - \frac{1}{B} = -\frac{R_2}{R_1} \tag{6}$$

respectively, while the error factor for both transfer functions remains the same.

By substituting Eqs. (2) into (3) and (4), the transfer functions of the feedback amplifiers in Fig. 1a and b can be re-expressed as

$$\frac{v_o(s)}{v_i(s)} = \frac{1}{B} \times \left(\frac{B\omega_{GBW}\omega_{np}}{s^2 + \omega_{np}s + B\omega_{GBW}\omega_{np}} \right) \tag{7}$$

and

$$\frac{v_o(s)}{v_i(s)} = \left(1 - \frac{1}{B}\right) \times \left(\frac{B\omega_{GBW}\omega_{np}}{s^2 + \omega_{np}s + B\omega_{GBW}\omega_{np}} \right). \tag{8}$$

Rearranging Eqs. (7) and (8), we obtain transfer functions of the feedback amplifiers, denoted as

$$\frac{v_o(s)}{v_i(s)} = \frac{1}{B} \times \left(\frac{\omega_o^2}{s^2 + \frac{\omega_o}{Q}s + \omega_o^2} \right) \tag{9}$$

and

$$\frac{v_o(s)}{v_i(s)} = \left(1 - \frac{1}{B}\right) \times \left(\frac{\omega_o^2}{s^2 + \frac{\omega_o}{Q}s + \omega_o^2} \right) \tag{10}$$

respectively. The quality factor Q and the natural frequency ω_o are given by

$$Q = \sqrt{\frac{B\omega_{GBW}}{\omega_{np}}} \quad \text{and} \quad \omega_o = \sqrt{B\omega_{GBW}\omega_{np}} \tag{11}$$

respectively. Referring to Eq. (11) in [10], a function of op amp's phase margin, denoted as ϕ_M, can be re-expressed as

$$\phi_M = \tan^{-1}\sqrt{\frac{2}{\sqrt{1 + 4\left(\frac{\omega_{GBW}}{\omega_{np}}\right)^2} - 1}}. \tag{12}$$

Substituting the quality factor Q Eqs. (11) into (12) and rearranging the result, we express the quality factor Q as

$$Q = \left[\frac{B^2}{4}\left(\frac{2}{\tan^2\phi_M} + 1\right)^2\right]^{\frac{1}{4}}. \tag{13}$$

According to Eq. (13), the plot of the quality factor Q versus the feedback factor B illustrated in Fig. 3, where it can be observed that for $\phi_M = 65°$ when B is less than unity, the values of the quality factor Q are less than 0.707. Consequently, the frequency response would not be maximally flat.

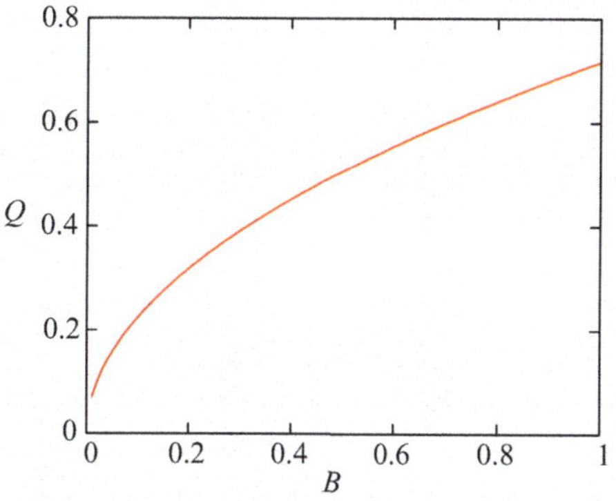

Fig. 3 Plot of Eq. (13) for $\phi_M = 65°$

3 Proposed Design Step of Op Amp Based Feedback Amplifier

If the op amp employed in the inverting amplifier in Fig. 1b is a two-stage CMOS op amp as depicted in Fig. 4, the conditions for the quality factor Q to be 0.707 are described by

$$g_{m1} = \frac{\omega_{-3dB}^2 C_C \left(1 + \frac{R_2}{R_1}\right)}{\frac{\omega_{-3dB}}{Q} - \frac{1}{C_L(R_1+R_2)}}, \tag{14}$$

$$\omega_{T6} = \frac{g_{m6}}{C_{gs6}} = \left(\sqrt{2}\omega_{-3dB} - \frac{1}{(R_1+R_2)C_L}\right)\frac{C_L}{C_C} \tag{15}$$

and

$$g_{m9} \cong \sqrt{2}\omega_{-3dB} C_L - \frac{1}{R_1+R_2} \tag{16}$$

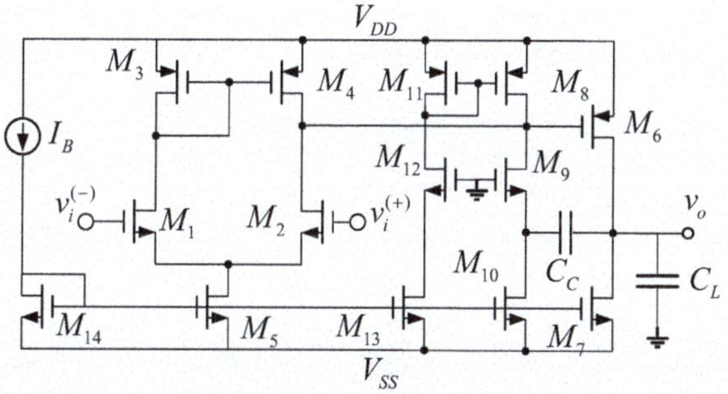

Fig. 4 Internal structure of the CMOS op amp

where g_{m1} and g_{m9} represent the transconductance of M_1 and M_9, respectively. Additionally, ω_{T6} is the transition frequency of M_6, and ω_{-3dB} is the frequency response at -3 dB. By substituting Eq. (11) and $I_{D1} = I_{D5}/2 = C_C SR/2$ into

$$g_m = \sqrt{2\mu_{n,p}C_{ox}\left(\frac{W}{L}\right)I_D} \tag{17}$$

and solving for $(W/L)_{1,2}$, we obtain

$$\left(\frac{W}{L}\right)_{1,2} = \frac{\omega_{-3dB}^4 C_C\left(1+\frac{R_2}{R_1}\right)^2}{\mu_n C_{ox} SR\left(\frac{\omega_{-3dB}}{Q} - \frac{1}{C_L(R_1+R_2)}\right)^2}. \tag{18}$$

where SR is the slew rate. Combining Eq. (15) with $\omega_T = ((2/3)\mu_p V_{eff})/L^2$ [15] and $V_{\text{eff}} = V_{HR}^{out+}$ yields

$$L_6 = \sqrt{\frac{3\mu_p V_{\text{HR}}^{\text{out}+}}{2\left(\sqrt{2}\omega_{-3dB} - \frac{1}{(R_1+R_2)C_L}\right)\frac{C_L}{C_C}}} \tag{19}$$

where V_{HR}^{out+} is the positive headroom voltage at output of the op amp. Substituting Eqs. (17) into (16) and solving for $(W/L)_9$ gives

$$\left(\frac{W}{L}\right)_9 \cong \frac{\left(\sqrt{2}\omega_{-3dB}C_L - \frac{1}{R_1+R_2}\right)^2}{2\mu_n C_{ox} I_{D9}}. \tag{20}$$

By integrating the above equations and incorporating part of the op amp design step for the open-loop amplifier outlined in [14], the design process for the inverting amplifier based on the CMOS op amp in Fig. 4 is summarized in Table 1.

4 Simulation Results

Using 0.35 μm AMS CMOS technology, inverted CMOS amplifiers based on op amps employing the CMOS configuration in Fig. 4 were designed. Two cases were considered: Case 1 with a DC gain of -2 ($R_1 = 100$ kΩ and $R_2 = 200$ kΩ) and a bandwidth of 10 MHz, and Case 2 with a DC gain of -5 ($R_1 = 100$ kΩ and $R_2 = 500$ kΩ) and a bandwidth of 4 MHz. The specification for both cases are detailed in Table 2. The amplifiers were designed using the approach outlined in [14], which is rooted in the traditional design strategy [6, 7, 11–14], setting the phase margin and unity-gain frequency at 65° and 20 MHz, respectively. Additionally, the proposed

Table 1 Design step for the inverting amplifier based on CMOS op amp

Step	Details
1	$I_{D7} = SR(C_C + C_L)$
2	$I_{D5} = C_C SR$
3	$\left(\frac{W}{L}\right)_{1,2} = \frac{\omega^4_{-3dB} C_C \left(1+\frac{R_2}{R_1}\right)^2}{\mu_n C_{ox} SR \left(\frac{\omega_{-3dB}}{Q} - \frac{1}{C_L(R_1+R_2)}\right)^2}$
4	$L_6 = \sqrt{\frac{3\mu_p V_{\mathrm{HR}}^{\mathrm{out}+}}{2\left(\sqrt{2}\omega_{-3dB} - \frac{1}{(R_1+R_2)C_L}\right)\frac{C_L}{C_C}}}$
5	$W_6 = \frac{2SR(C_C+C_L)}{\mu_p C_{ox}\left(V_{HR}^{Out+}\right)^2} L_6$
6	$(W/L)_{5,14} = \frac{2SRC_C}{\mu_n C_{ox}(V_{\mathrm{HR}}^{\mathrm{CM}-} - V_{th} - V_x)^2}$ where $V_x = \frac{SR}{\omega^2_{-3dB}\left(\frac{\omega_{-3dB}}{Q} - \frac{1}{C_L(R_1+R_2)}\right)\left(1+\frac{R_2}{R_1}\right)}$
7	$(W/L)_7 = \left(\frac{C_C+C_L}{C_C}\right)(W/L)_{5,14}$
8	$(W/L)_{3,4} = \frac{(W/L)_6}{2(W/L)_7}(W/L)_{5,14}$
9	$\left(\frac{W}{L}\right)_9 \cong \frac{\left(\sqrt{2}\omega_{-3dB} C_L - \frac{1}{R_1+R_2}\right)^2}{2\mu_n C_{ox} I_{D9}}$

Table 2 Inverting amplifier specification

Supply voltages		±1.5 V
Slew rate: SR		10 V/μsec
Load capacitance: C_L		2 pF
Quality factor: Q		0.707
Case 1	DC gain	−2
	−3 dB bandwidth (MHz)	10 MHz
Case 2	DC gain	−5
	−3 dB bandwidth (MHz)	4 MHz

design step from Table 1. was incorporated. Design parameters for Case 1 and Case 2 are presented in Tables 3 and 4, respectively. As illustrated in Figs. 5, 6, 7 and 8, the simulated frequency and transient responses of the inverting amplifiers obtained from both design steps are similar. However, it can be observed from Tables 5 and 6 that in Cases 1 and 2, power consumption of the amplifiers obtained from the proposed design step is respectively 32.09 and 45.42% less than that of the amplifiers obtained from the step in [14].

Table 3 Design parameters case1

Parameter		Traditional	Proposed
Supply voltages (V)		± 1.5	
C_C (pF)		1	
C_L (pF)		2	
R_1 (kΩ)		100	
R_2 (kΩ)		200	
I_B (μA)		14	12.5
W/L (μm/μm)	$M_{1,2}$	25/1	25/1
	$M_{3,4}$	5.85/1	5.85/1
	$M_{5,14}$	2.15/5	2.15/5
	M_6	60/2.25	139.7/4
	M_7	2.6/2	2.55/2
	$M_{8,11}$	10/1	10/1
	$M_{9,12}$	25/0.5	6.55/0.5
	$M_{10,13}$	4.2/5	2.15/5

Table 4 Design parameters case 2

Parameter		Traditional	Proposed
Supply voltages (V)		± 1.5	
C_C (pF)		1	
C_L (pF)		2	
R_1 (kΩ)		100	
R_2 (kΩ)		500	
I_B (μA)		14	10
W/L (μm/μm)	$M_{1,2}$	25/1	27.9/1
	$M_{3,4}$	5.85/1	5.85/1
	$M_{5,14}$	2.15/5	2.25/5
	M_6	60/2.25	221.4/6.3
	M_7	2.6/2	27/2
	$M_{8,11}$	10/1	10/1
	$M_{9,12}$	25/0.5	1.05/0.5
	$M_{10,13}$	4.2/5	2.25/5

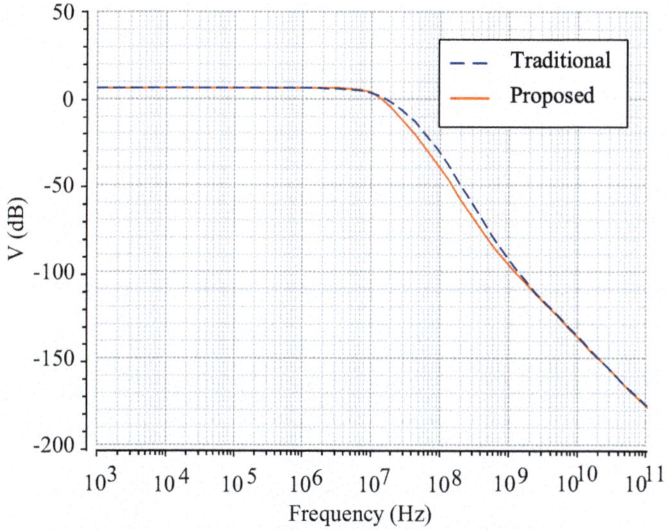

Fig. 5 The closed-loop frequency response of the amplifiers case 1

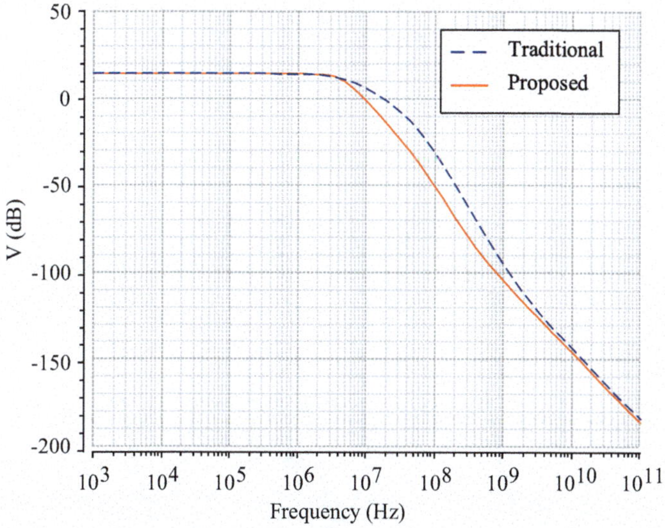

Fig. 6 The closed-loop frequency response of the amplifiers case 2

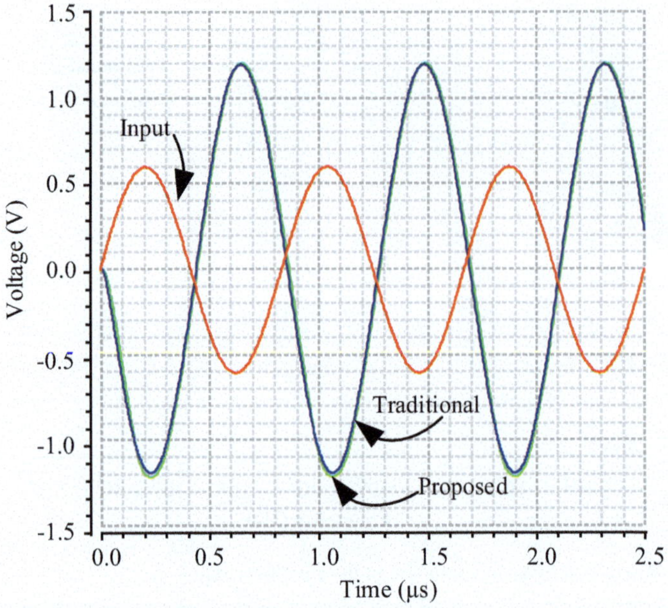

Fig. 7 The transient response of the amplifier in Case 1 involves applying a sinusoidal wave with a peak voltage of 0.6 V at 1.2 MHz to the input

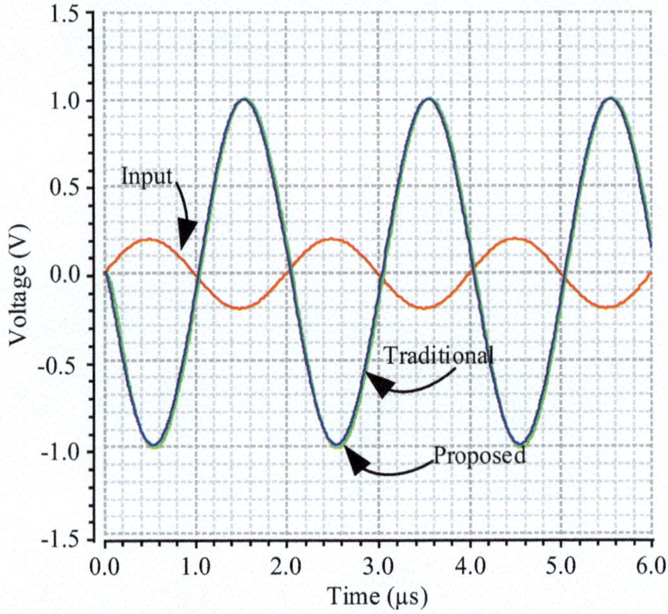

Fig. 8 The transient response of the amplifier in Case 2 involves applying a sinusoidal wave with a peak voltage of 0.2 V at 500 kHz to the input

Table 5 Simulation results of the inverting amplifier case 1

Parameter	Traditional	Proposed
Closed-loop gain	1.998	1.999
−3 dB bandwidth (MHz)	10.36	10.2
Quality factor: Q	0.515	0.714
Slew rate: SR (V/μs)	13.66	12.05
Power (μW)	375.5	255
Output noise (μV_{RMS}) (1 Hz–10 MHz)	357.1	381.7

Table 6 Simulation results of the inverting amplifier case 2

Parameter	Traditional	Proposed
Closed-loop gain	4.995	4.997
−3 dB bandwidth (MHz)	4.39	4.06
Quality factor: Q	0.516	0.706
Slew rate: SR (V/μs)	12.32	9.98
Power (μW)	375.5	204.96
Output noise (μV_{RMS}) (1 Hz–10 MHz)	608.8	593.6

5 Conclusion

This paper has presented a comprehensive design approach for op-amp-based CMOS feedback amplifiers, showcasing the advantages of the proposed design steps over traditional methods. By designing feedback amplifiers with a gain bandwidth (GBW) of 20 MHz and a quality factor (Q) of 0.707 using 0.35 μm CMOS technology, simulations have demonstrated that the proposed design not only achieves a more flat frequency response but also significantly reduces power consumption compared to the traditional approach. These findings emphasize the effectiveness of the proposed design strategy in optimizing amplifier performance. As technology continues to advance, incorporating such design steps becomes crucial for achieving efficient and high-performance integrated circuits.

References

1. Ahuja BK (1983) An improved frequency compensation technique for CMOS operational amplifiers. IEEE J Solid-State Circuits SC-18(6):629–633. (December)
2. Fortier GJ, Filanovsky IM (1987) A compensation technique for two-stage CMOS operational amplifier. Microelectron J 18(3):50–57. (May–June)
3. Johns DA, Martin K (1997) Analog integrated circuit design. Wiley, New York
4. Palmisano G, Palumbo G (1997) A compensation strategy for two-stage CMOS opamp based on current buffer. IEEE Trans Circuits Syst 44(3):257–262. (March)

5. Razavi B (2000) Design of analog CMOS integrated circuits. McGraw-Hill Companies
6. Mahattanakul J, Chutichatuporn J (2005) Design procedure for two-stage CMOS opmap with flexible noise-power balancing scheme. IEEE Trans Circuits Syst 52(5):1508–1514. (August)
7. Mahattanakul J (2005) Design procedure for two-stage CMOS operational amplifiers employing current buffer. IEEE Trans Circuits Syst 52(11):766–769. (November)
8. Amoroso FA, Pugliese A, Cappuccino G (2009) Design considerations for fast-settling two-stage miller-compensated operational amplifiers. In: The 16th international conferences on electronics, circuits and systems (ICECS 2009), pp 5–8
9. Pugliese A, Amoroso FA, Cappuccino G, Cocorullo G (2010) Analysis of op-amp phase margin impact on SC Σ Δ modulator performance. Microelectron J 41(7):440–446
10. Chanapromma C, Mahattanakul J (2021) New method to design feedback amplifier employing two-stage CMOS op amp. Int J Electron Commun (AEU) 132:153616
11. Allen PE, Holberg DR (2002) CMOS analog circuit design, 2nd edn. Oxford University Press, New York
12. Palmisano G, Palumbo G, Pennisi S (2001) Design procedure for two-stage CMOS tramsconductance operational amplifiers: a tutorial. Analog Integr Circ Sig Process 27:179–189
13. Chanapromma C, Mahattanakul J (2019) Redesign procedure for two-stage CMOS op amp with least error of frequency response. In: The 16th international conferences on electrical engineering/electronics, computer, telecommunications and information technology (ECTI-CON 2019), pp 717–720
14. Chanapromma C, Mahattanakul J (2020) Improved design procedure for two-stage CMOS op amp employing current buffer. In: The 17th international conferences on electrical engineering/electronics, computer, telecommunications and information technology (ECTI-CON 2020), pp 384–387
15. Gray PR, Hurst PJ, Lewis SH, Meyer RG (2001) Analysis and design of analog integrated circuits, 4th edn. John Wiley & Sons, Inc.

Dual-Edge Triggered Reset Synchronizer for I2C Protocol

Myint-Wai Phyu, Joscel Ann Agabao Ignacio, and Richard Oey

Abstract In this paper, dual-edge triggered reset synchronizer for Inter Integrated Circuit (I2C) protocol is proposed. The proposed reset synchronizer design aims for the asynchronous reset to have asynchronous assertion and synchronous deassertion for I2C slave protocol, in which system reset and software reset is synchronized to a targeted clock. This is to avoid the metastability issue when the reset is released with respect to the target clock. Taking the role of I2C START condition, which signifies the beginning of a transmission, the proposed methodology uses START condition as a clock to synchronize the external reset or software reset before transmission begins. Both edges of the START signal will be used for synchronization. Since START timing must be respected according to I2C protocol, it is safe to be used as a clock without impacting any setup/hold (reset/recovery) timing issue.

Keywords Reset synchronizer · I2C · START · Dual-edge triggered · UVM

1 Introduction

Asynchronous resets are traditionally employed in VLSI designs for bringing synchronous circuitry to a known state after power up. Asynchronous reset release operation must be coordinated with the synchronous logic clock signal to eliminate synchronization failures due to possible contention between the reset and the clock [1]. A lack of such coordination leads to intermittent failures on power up. Reset may be either synchronous or asynchronous relative to the clock signal. The benefits

M.-W. Phyu (✉) · J. A. A. Ignacio · R. Oey
STMicroelectronics, Singapore, Singapore
e-mail: myint-wai.phyu@st.com

J. A. A. Ignacio
e-mail: joscel-ann-agabao.ignacio@st.com

R. Oey
e-mail: richard.oey@st.com

M. Ma (ed.), *Proceedings of the 12th International Conference on Communications, Circuits, and Systems*, Lecture Notes in Electrical Engineering 1193,
https://doi.org/10.1007/978-981-97-2636-3_12

of using synchronous reset are that they are deterministic and do not incur metastability. However, they require an active clock edge and may impact the timing of the data paths. In the case of an asynchronous reset, the converse is true: they do not require an active clock to the flip-flops to a known state, they have lower latency in comparison, and can exploit special flip-flop input pins that do not impact data path timing. However, they may also cause metastability and non-deterministic behavior in flip-flops. Asynchronous resets are widely employed in digital designs. In many cases, asynchronous resets can be replaced by synchronous ones, but there are some cases in which the asynchronous reset functionality is compulsory. One example is a synchronous design that gets no active clock at power up (the clock is either unstable or gated for power reduction), but requires a certain known state for its external interfaces [2]. Another example is interface protocol slave design, like that of I2C or SPI, that do not have an active clock before communication from the master starts. In this case, the resets of the slaves are asynchronous to the external communication clocks, thus making the reset mechanism not entirely straightforward. Although the relative timing between clock and reset can be ignored during reset assertion, the reset release must be synchronized to the clock. Reset release edge synchronization is necessary to avoid potential metastability [3, 4].

Figure 1 shows the state-of-the-art reset synchronizer. It consists of two registers connected in series, the input of the first register is tied to VDD. The asynchronous reset signal is connected to the reset pin of both registers. The output of this circuit has a synchronized de-assertion relative to the clock input.

When the reset is asserted, it first propagates to reset synchronizer flops. It resets both the flops of reset synchronizer asynchronously (without waiting for clock edge), thereby generating reset assertion for fanout registers. Similarly, the deassertion of reset first reaches the two flops of reset synchronizer.

In Fig. 2, the first flop in chain propagates 1 to the output upon arrival of a clock edge. Upon next clock edge, the reset de-assertion timing (recovery and removal checks timing) should be met from second stage of reset synchronizer as the deassertion is synchronous. With this implementation, two clock edges is required for synchronous reset deassertion. Therefore, this paper proposes a dual-edge triggered

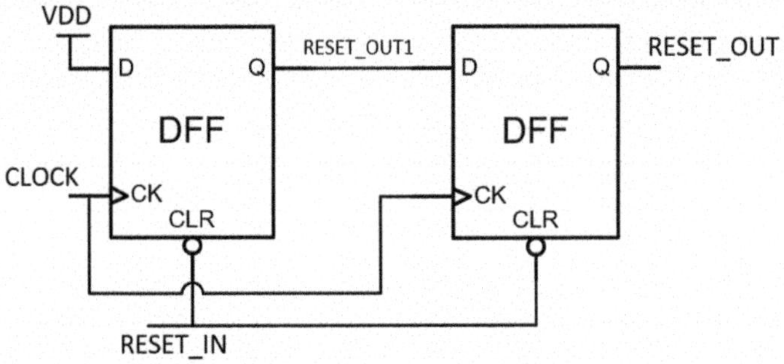

Fig. 1 State-of-the-art reset synchronizer

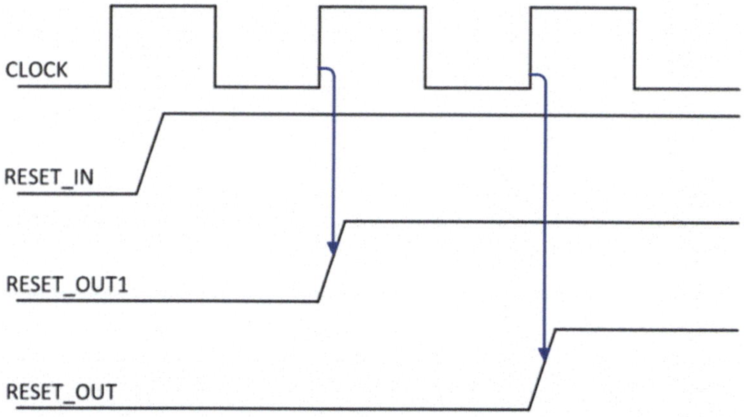

Fig. 2 Reset synchronizer timing

reset synchronizer for I2C protocol which aims to start the communication as soon as an active clock is received without impacting de-assertion timing.

The paper is organized as follows. Section 2 explains the I2C protocol and its timing. Section 3 presents the proposed dual-edge triggered reset synchronizer and the detailed implementation of the proposed design. Section 4 explains the simulation results. Finally, Sect. 5 concludes the paper.

2 I2C Protocol

I2C stands for Inter-Integrated Circuit. It is a bus interface connection protocol incorporated into devices for serial communication. Two wires, serial data (SDA) and serial clock (SCL), carry information between the devices connected to the bus. A master is the device which initiates a data transfer on the bus and generates the clock signals to permit that transfer. At that time, any device addressed is considered a slave. All transactions begin with a START (S) and are terminated by a STOP (P) (see Fig. 3). A HIGH to LOW transition on the SDA line while SCL is HIGH defines a START condition. A LOW to HIGH transition on the SDA line while SCL is HIGH defines a STOP condition [5].

START and STOP conditions are always generated by the master. The bus is considered to be busy after the START condition. The bus is considered to be free again a certain time after the STOP condition. The I2C data transfer format (START + ADDRESS + DATA + STOP) is shown in Fig. 4.

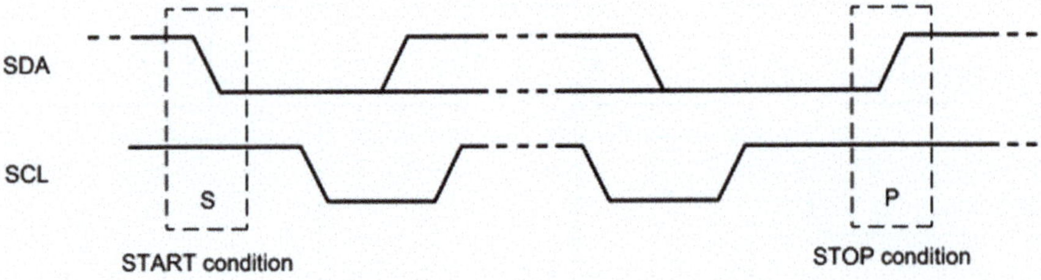

Fig. 3 START and STOP conditions

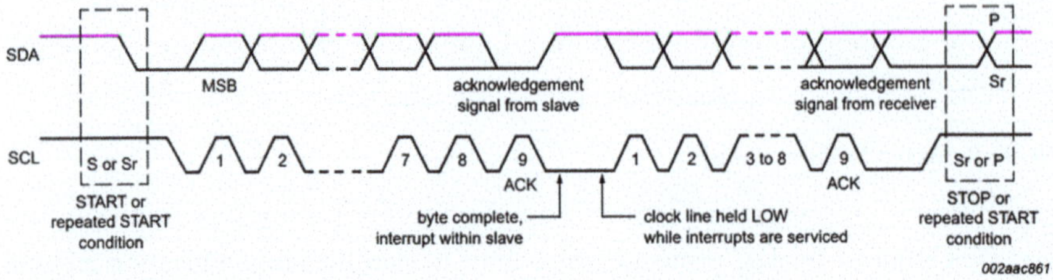

Fig. 4 Data transfer on the I2C bus

As seen in Fig. 4, once START is detected on the bus, SDA is sampled on active (rising) edge of SCL after the START condition. If reset is synchronized on SCL clock, data sampling starts after reset is ready and the first I2C device address sent may not be wholly received by the slave, acting as a 'dummy' byte that serves only to synchronously de-assert the reset with the target clock, the SCL (Fig. 4). Thus, with this sequence, slave performance may be impacted, as it will not response to the first address byte. Therefore, this paper introduced a dual edge triggered reset synchronizer without compromise the first device address, using the START signal as a clock.

3 Proposed Dual-Edge Triggered Reset Synchronizer

3.1 Proposed Reset Synchronizer

Figure 5 shows the schematic of the proposed dual-edge triggered reset synchronizer. As seen below, first flip-flop is on positive-edge and second flip-flip is on negative edge and clocked by START_PULSE. Figure 6 shows the START_PULSE generation based on the START condition. And START_PULSE is used as a clock to synchronize the external reset or software reset before transmission begins (Fig. 7).

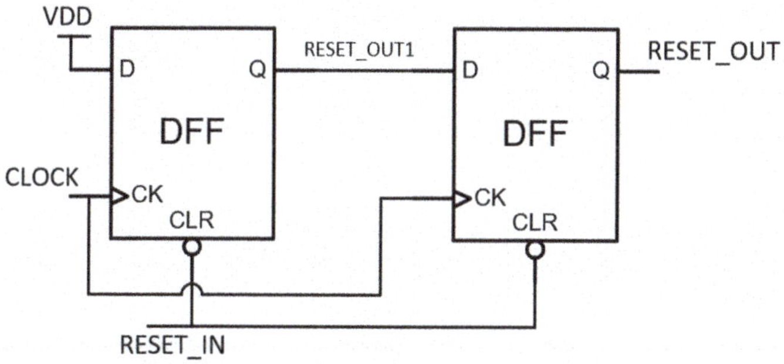

Fig. 5 Proposed dual-edge triggered reset synchronizer

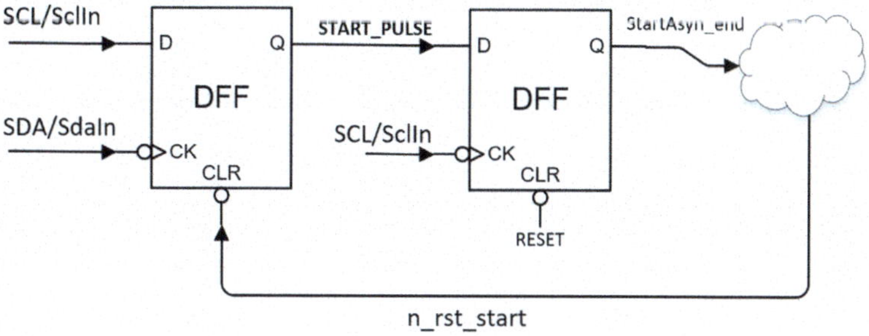

Fig. 6 Schematic of START_PULSE generation logic

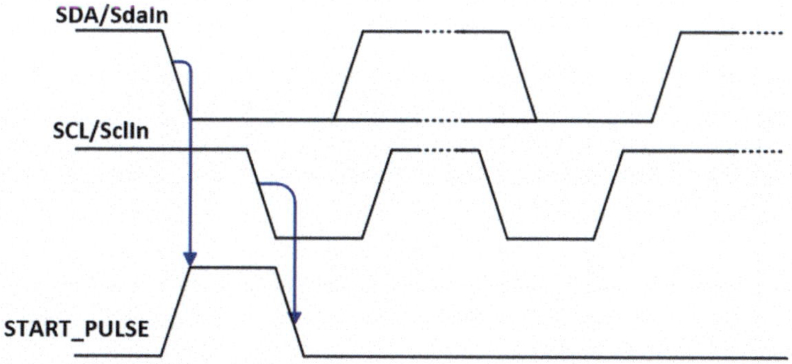

Fig. 7 START_PULSE signal

3.2 *Timing Analysis*

This section analyzes the metastability due to reset de-assertion relative to the clock edge on the proposed dual-edge triggered reset synchronizer. As per I2C protocol timing definition (Fig. 8), START condition timing ($t_{HD;STA}$) must be respected. The I2C-bus timing characteristics is given in Table 1.

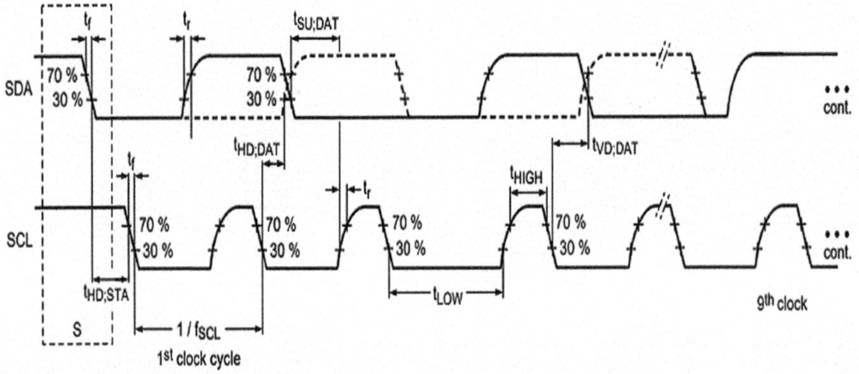

Fig. 8 Timing definition for the I2C-bus

Table 1 Characteristics of the SDA and SCL bus lines for fast and fast-mode plus I2C-bus devices

Symbol	Parameter	Fast-mode		Fast-mode plus	
		Min	*Max*	*Min*	*Max*
f_{SCL} (kHz)	SCL clock frequency	0	400	0	1000
$t_{HD;STA}$ (μs)	hold time (repeated) START condition	0.6	–	0.26	–
t_{LOW} (μs)	LOW period of the SCL clock	1.3	–	0.5	–
t_{HIGH} (μs)	LOW period of the SCL clock	0.6	–	0.26	–
$t_{SU;STA}$ (μs)	set-up time for a repeated START condition	0.6	–	0.26	–

In the flip-flop design (Fig. 5), reset has put a priority on whatever occurs on inputs when activated. In this reset synchronizer, D input of DFF1 is connected to a static signal to avoid any metastable state. Due to the use of both edges, only concern is the delay between 1st DFF and 2nd DFF and arrival of reset relative to the clock edge. As illustrated in Fig. 9, START_PUSE (red box) always follows the minimum $t_{HD;STA}$ and it does not depend on PVT (process, voltage, temperature) conditions. Therefore, there will be no issue of the delay between 1st DFF and 2nd DFF. The issue to be focused is reset arrival with respect to the clock edge. Following this, 3 usecases will be emphasized:

Case I: reset is de-asserted before START.
Case II: reset de-asserts at the same time as START.
Case III: reset de-asserts after START.

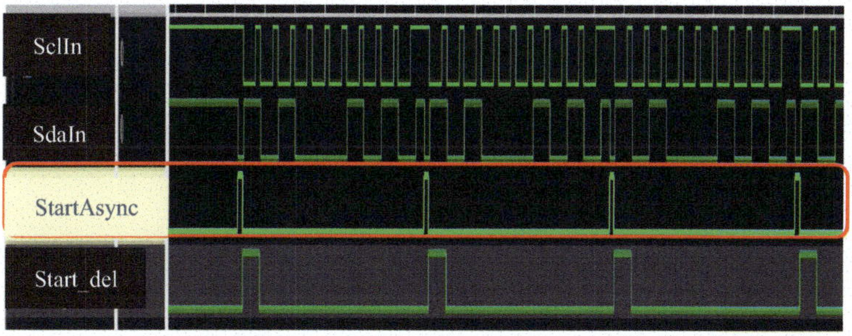

Fig. 9 START signal simulation

3.3 Case I

In this case, reset is de-asserted before START condition arrived. D is seen at '1' before rising edge of RESET_IN and '1' is propagated to Q/RESET_OUT1 on rising edge of START_PULSE and then Q/RESET_OUT goes to '1' on falling edge of START_PULSE (Fig. 10).

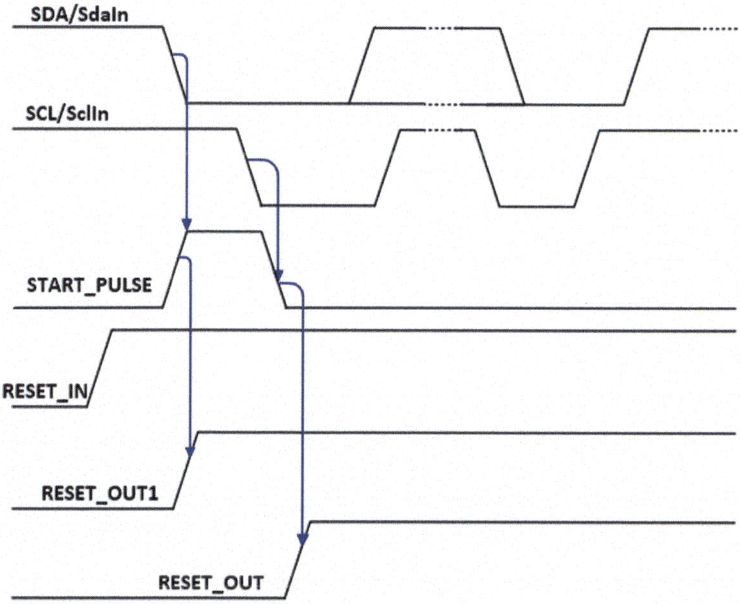

Fig. 10 Reset de-assertion before START

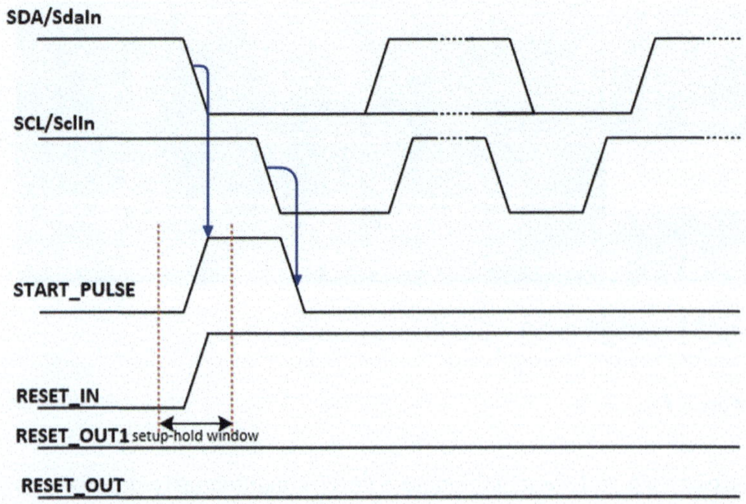

Fig. 11 Reset de-assertion at the same time as START

3.4 Case II

In this case, reset is de-asserted at the same time as START condition within reset recover/removal (setup/hold) interval. D is seen at '1' may not be sampled by 1st DFF and result is unpredictable but 2nd DFF Q is stable and it can have one START_PULSE incertitude (Fig. 11). In this scenario, RESET_OUT will stay in reset state, which in turn 1st device address will not be processed and slave will send a NACK to mater. Master I2C will initiate another transaction and reset will properly sampled on next START.

3.5 Case III

In this case, reset is de-asserted after START condition and timing is met. As seen in Fig. 12, 1st START is missed and D input at '1' is sampled on next START.

4 Simulation Testbench and Waveforms

The proposed reset synchronizer has been verified in Universal Verification Methodology (UVM) simulation. UVM testbench illustrated Fig. 13 consists of AHB VIP, AHBtoAPB bride for register access, I3C VIP for master I2C and IOs for SDA and SCL and I2C slave DUT.

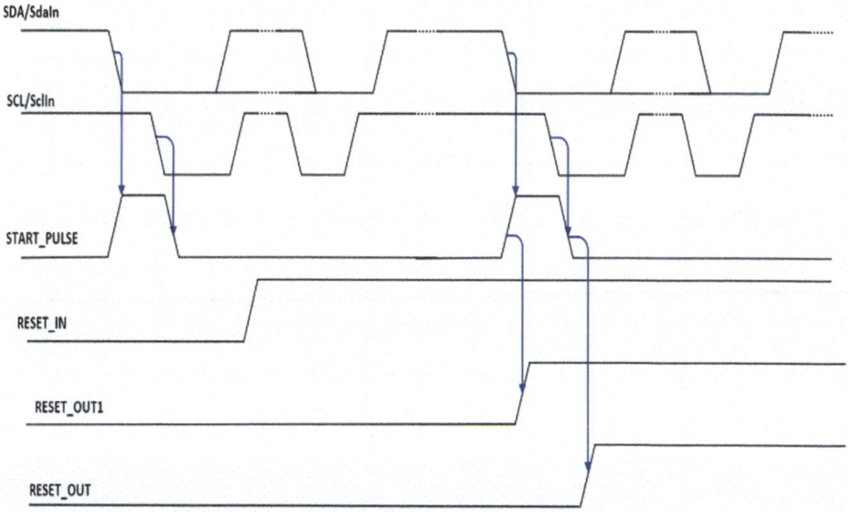

Fig. 12 Reset de-assertion after START

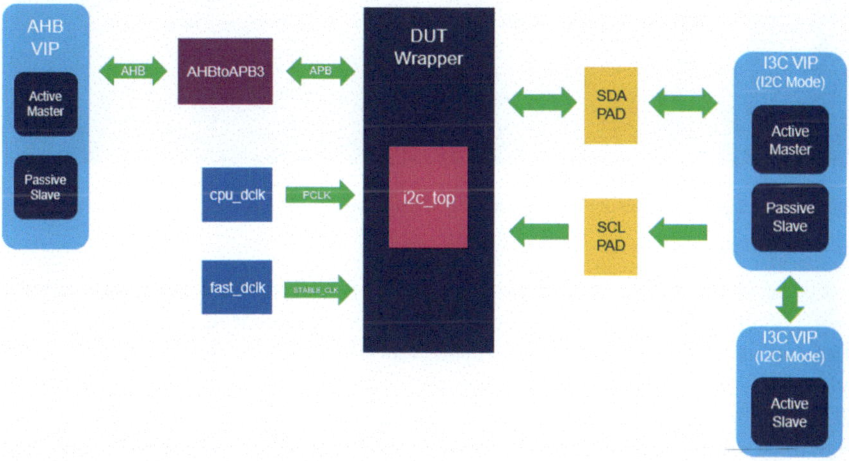

Fig. 13 UVM testbench

The first simulation (Fig. 14) shows reset de-assertion before START condition sent by master I2C. In this simulation, the output of the 2nd DFF (synchronized reset) is stable after falling edge of START_PULSE (red box). Internal counter starts counting at every SCL (SclIn) falling edge and shifter starts receiving data on SDA (SdaIn) line on every rising edge of SCL (SclIn).

The second simulation (Fig. 15) shows asynchronous reset assertion while communication is on-going. It can be seen from simulation that asynchronous reset assertion is effective immediately and all internal flip-flops goes to reset state.

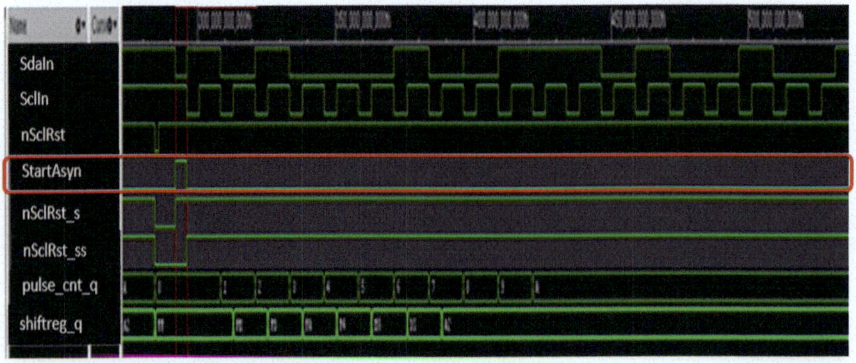

Fig. 14 Reset de-assertion before START

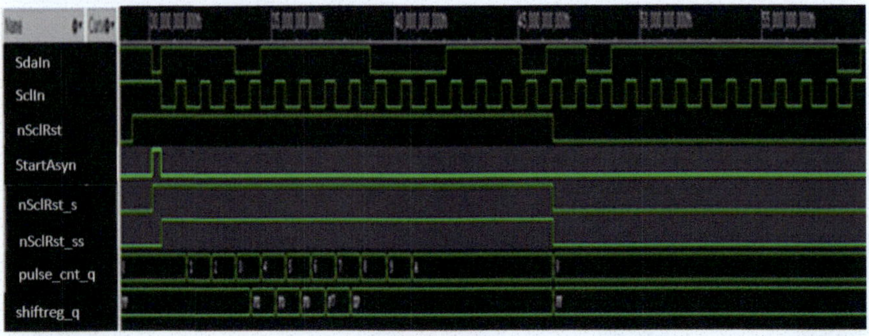

Fig. 15 Reset assertion during communication is on-going

Figure 16 shows that nSclRst is synchronized to SCL domain using conventional reset synchronizer, discussed in Sect. 1. It can be seen that nSclRst_ss is latched after 2nd rising edge of SclIn. As a result, 1st SclIn pulse is missed and 1st address byte is used as a dummy byte for reset to synchronize on SCL domain. Thanks to I2C START condition which is used as a clock to resync on SCL domain using dual edge triggered reset synchronizer. With this special feature of I2C START and dual edge triggered reset synchronizer, dummy byte is not required and thus it improved in terms of performance.

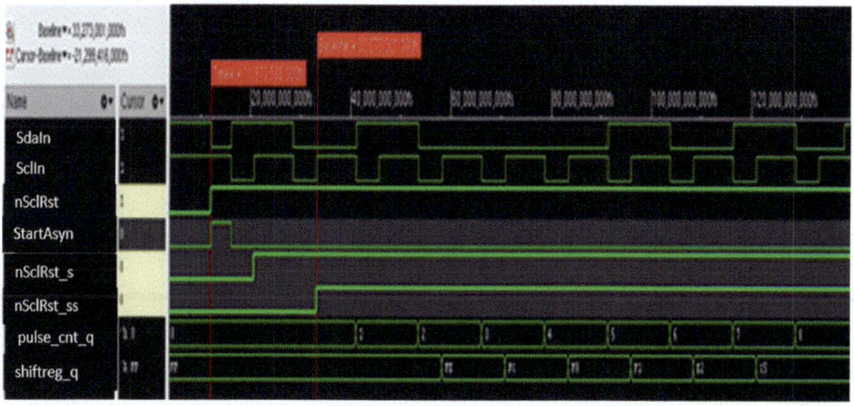

Fig. 16 Conventional reset synchronizer

5 Conclusion

Reset architecture of a digital design can be quite complex. Typically, System-on-Chips (SoC) have multiple sources of reset, such as power-on-reset, hardware reset and software reset. These multiple reset domains make the design potentially exposed to metastability issues due to possible contention between the reset and the clock. The simulation results showed that the proposed reset synchronizer design performs both asynchronous reset to have asynchronous assertion and synchronous deassertion for I2C slave protocol.

References

1. Ahmed, Nouh K, Abbas A (2017) Multiple reset domains verification using assertion based verification. In: 2017 IFIP/IEEE International conference on very large scale integration (VLSI-SoC), Abu Dhabi, 2017, pp 1–6
2. Kasim M, Gupta V, Jebin M (2020) Methodology for detection glitch on clock, reset and CDC path. In: Proceedings of the fifth international conference on communication and electronics systems (ICCES 2020) pp 300–304
3. Jones J, Yang S, Greenstreet M (2009) Synchronizer behavior and analysis. In: Proc. IEEE international symposium on circuits and systems (ASYNC 09), IEEE CS Press, 2009, pp 117–126
4. Fawzy M, Elgohary A, Ibrahim H (2020) Noise reduction in reset domain crossings verification using formal versification. IEEE east-west design & test symposium EWDTS 2020, pp 1–5
5. UM10204 I2C-bus specification and user manual Rev. 6–4 April 2014

Design of Quality High Factor Fully Differential CMOS Active Inductor with Cascode Current Mirror for 2–5 GHz Frequency Ranges

Hussein Anes AlShaikh Ali, Sohiful Anuar Zainol Murad, Faizah Abu Bakar, Ahmad Fariz Hasan, and Jamilah Karim

Abstract This paper presents a fully differential CMOS active inductor with a cascoded current mirror for 2–5 GHz frequency ranges applications. The proposed active inductor is designed using CMOS 0.18-μm process. The circuit is realized by a fully differential cross-coupled pair of transistors to provide negative feedback for high-quality factor (Q). The cascoded current mirror is employed as a biasing circuit for controlling the current source to vary the negative feedback and tuning the Q factor. Meanwhile, the two resistors at the differential structure help to control the frequency ranges. According to the simulation results, the higher Q factor of 12,580 k with an inductance of 11.4 nH is achieved at 3.6 GHz frequency. The frequency range from 2.6 GHz to 4.4 GHz can be obtained by tuning the value of the resistor. Moreover, it is observed that the supply voltage and current source can be varied to tune the Q factor values from 100 to 900 k. It is proved through a performance comparison with previously published research that the suggested active inductor is appropriate for RF in gigahertz frequency range applications.

H. A. A. Ali
Faculty of Electronic Engineering Technology (FTKEN), Universiti Malaysia Perlis (UniMAP), Arau, Perlis, Malaysia
e-mail: husseinanes@studentmail.unimap.edu.my

S. A. Z. Murad (✉) · F. A. Bakar · A. F. Hasan
Center of Excellent for Micro System Technology (MiCTEC), Faculty of Electronic Engineering Technology, Universiti Malaysia Perlis (UniMAP), Arau, Perlis, Malaysia
e-mail: sohiful@unimap.edu.my

F. A. Bakar
e-mail: faizah@unimap.edu.my

A. F. Hasan
e-mail: farizhasan@unimap.edu.my

J. Karim
School of Electrical Engineering, College of Engineering, Universiti Teknologi MARA (UiTM), Shah Alam, Selangor, Malaysia
e-mail: jamilah329@uitm.edu.my

M. Ma (ed.), *Proceedings of the 12th International Conference on Communications, Circuits, and Systems*, Lecture Notes in Electrical Engineering 1193,
https://doi.org/10.1007/978-981-97-2636-3_13

Keywords CMOS · Fully differential · Active inductor · Quality factor · Rf front-end

1 Introduction

Inductors are an important part of the circuit, that works mainly to block high-frequency signals and resist the change in current in their simpler form, caused by the fluctuations of the signal [1]. For years, spiral inductors have been the most reliable and the most well-known tried and true method in RF front-end design [2, 3]. It is a form of wire or coil that stores the energy as a magnetic field, which delays the voltage by 90° when the current passes through it, acting as a passive component [4]. However, with further developments and especially in smaller IC designs, the limitations that come from those devices are more commonly shown. Whether it was the low quality-factor factor (Q), larger die area, or wiring substrate losses [5]. With the continuous shrinking of devices technology, designing them with as little losses as possible is essential, while decreasing the chip area even further.

The gyrator was first proposed by Bernard D.H. Tellegen of Philips Research Laboratory as a circuit of two ports, which would be later developed as a Gyrator-C as the bases of the active inductor (AI). The AI has a much higher Q and a wider range of inductance, while also being insensitive to electromagnetic interference [6–10]. It is a circuit that mimics the behavior of the inductor while being much smaller and having lower losses. The problem with AI is the poor linearity and high-power consumption, compared to no power at all from the spiral inductor [11]. Which is the focus in designing AI, to balance its issues while improving the pros of the device.

This paper proposes a high-Q factor fully differential AI with a cascoded current mirror. The proposed AI is designed utilizing CMOS 0.18-μm technology. The proposed AI circuit design is discussed in Sect. 2. The simulation results are investigated in Sect. 3 and the conclusion is given in Sect. 4.

2 Proposed Circuit Design

A fully differential cross-coupled pair of transistors realizes the suggested active inductor (AI), which provides negative feedback for high-quality factor (Q). The cascoded current mirror is employed as a biasing circuit for controlling the current source to vary the negative feedback and tuning the Q factor. A cascoded current mirror has the advantages of giving much higher output impedance and much more consistent current values due to extinguishing the effects of channel length modulation [12]. The differential structure for the active inductor has many benefits such as better linearity and sensitivity [13]. This gives the device a wider array of parameters in order to reach the required results.

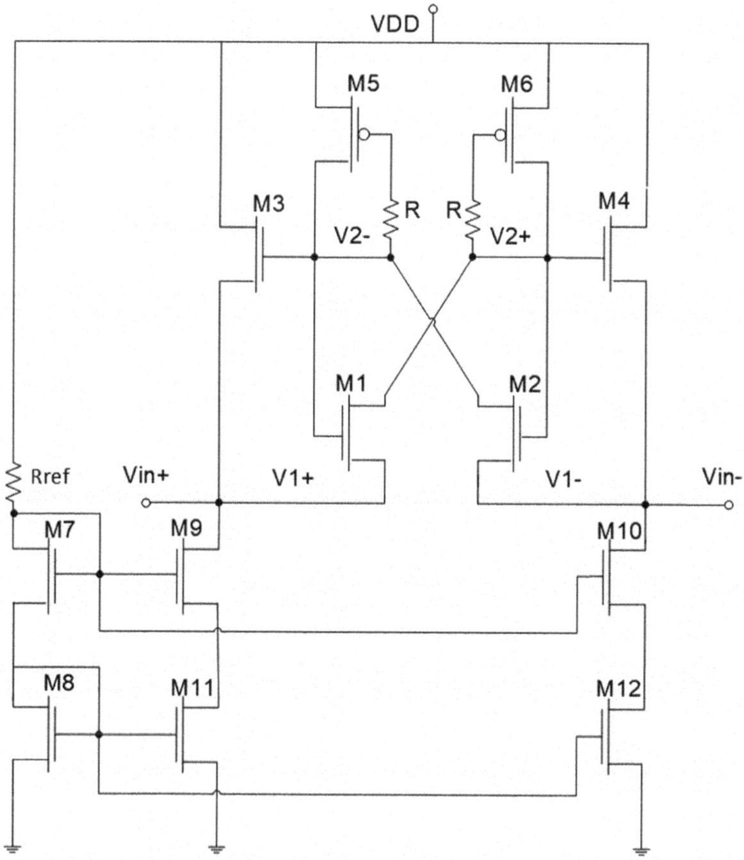

Fig. 1 The proposed fully differential active inductor with cascode current mirror

Figure 1 shows the proposed active inductor. With a 1.2 V supply voltage and 0.18-μm CMOS technology, the proposed circuit is created with Cadence Virtuoso software. The four primary NMOS transistors M1–M4 and the two PMOS transistors M5–M6 at the top make up the fundamental design of the proposed circuit. The NMOS device is connected through the drain, whereas the PMOS devices are connected through their source terminal. The M3, M4, M5, and M6 are directly connected to the supply voltage VDD. Each of the aforementioned MOSFETs has a drain that connects M1 and M2 to the gate. Transistors are all operating in saturation mode. When the differential input signal is applied at sources M1 and M2, the differential circuit functions. Increases in Vin + lead to increases in V2 + , and decreases in Vin- result in decreases in V2-. To guarantee equal variation for V1 + and V1-, M3 and M4 share a single source topology. Both sides get current from the cascaded current mirror of M7-M12 in the meantime, and the self-resonance frequency is adjusted by adding a resistor R. A high Q factor can be obtained by adjusting two parameters. In the first parameter, a resistance Rref can be tuned to increase the current source for M1/M3 and M2/M4, whereas in the second parameter the supply voltage VDD can be varied to increase the Q factor.

3 Simulation Results

The simulation was done with Cadence virtuoso, using CMOS 0.18 μm technology. Table 1 shows the dimensions for all of the CMOS transistors of the proposed fully differential AI. Two PMOS transistors M5 and M6 at the top are connected to four NMOS transistors (M1-M4). The transistor M7-M12 is a cascoded current mirror for biasing purposes.

Figure 2 displays the real and imaginary parts of the input impedance vs frequency. The active inductor's operational frequency range is 0–10 GHz, where the imaginary component of the input impedance is positive. It is evident that the proposed differential AI achieves the highest real resistance value of 400 Ω at 10 GHz and the highest imaginary reactance value of 300 Ω at 5 GHz. In the meantime, the frequency ranges of 0–4 GHz exhibit the lowest input resistance for the differential AI, suggesting that greater quality factors can be attained in these bands.

The Q factor can be tune by changing the Rref which the current source in the differential structure will be changed accordingly. The effects from Rref to quality

Table 1 Transistor dimension

Transistor	Dimension ((No. of finger × W)/L)
M1, M2	(1 × 3.6 μm)/0.18 μm
M3, M4	(1 × 3.0 μm)/0.18 μm
M5, M6	(5 × 1.2 μm)/0.18 μm
M7-M12	(1 × 1 μm)/0.18 μm

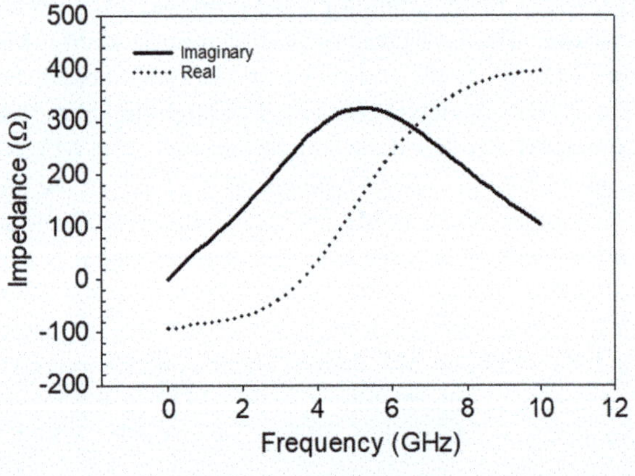

Fig. 2 The simulated frequency response of input impedance

factor at 3.5 GHz is shown in Fig. 3. As can be seen, the Q factor changes significantly from 40 to 500 k when the Rref is tuned from 5775 Ω to 5778 Ω. It's observed that the value of Rref is too sensitive to the Q factor, therefore the Rref need to be properly find tune to obtain the desired Q factor.

Figure 4 shows the effects of R on the self-resonance frequency. The self-resonance frequency of AI is made tuneable by incorporating the resistor R's value. In Fig. 4, the frequency of 2.8, 3.6, and 4.2 GHz can be achieved at the resistance of 1.3 kΩ, 1.7 kΩ, and 2.1 kΩ, respectively. It's observed that when the resistance is increased, the frequency is also increased.

Figure 5 depicts the simulated Q factor with difference VDD at 3.0 GHz. The Q factor from 50 to 900 k can be tuned by varies the VDD from 2.2 V to 2.7 V.

The variation of inductance with respect to the supply voltage VDD is illustrated in Fig. 6. The inductance values is decreased from 11.1 nH to 8.2 nH when the VDD increased from 2.2 V to 2.7 V. Finally, Fig. 7 shows the variation of inductance with respect to the frequency ranges, As can be observed, the inductances value is changing from 11 nH to 14.5 nH for the frequency ranges from 2.6 GHz to 4.4 GHz.

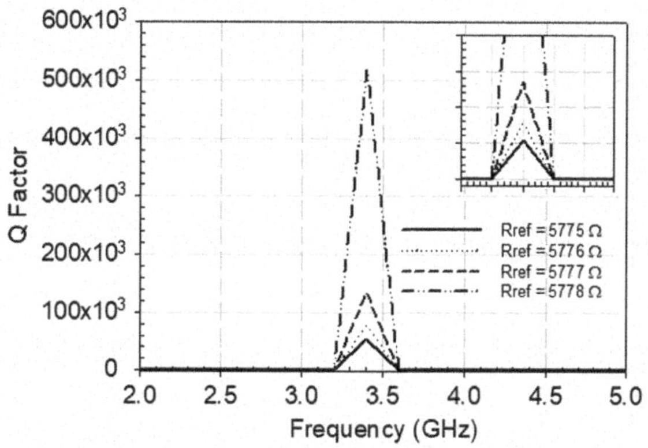

Fig. 3 The effects of Rref on the quality factor at 3.5 GHz

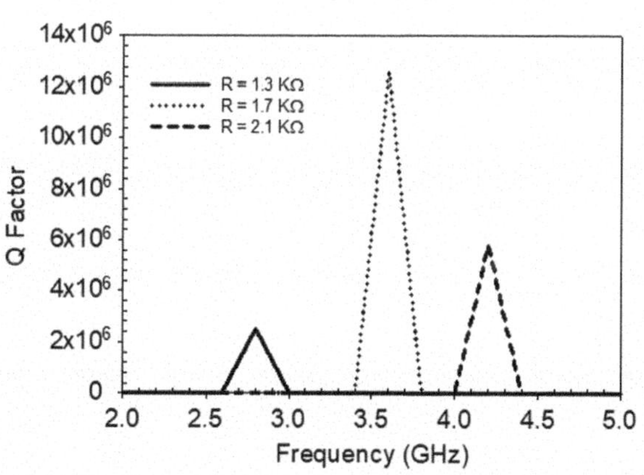

Fig. 4 The effects of R with self-resonance frequency and Q factor

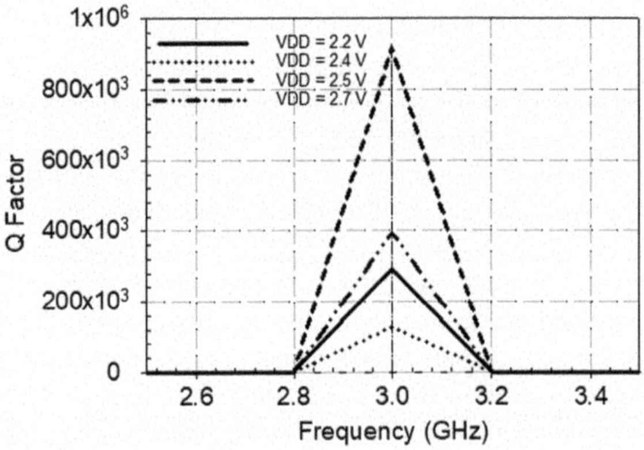

Fig. 5 The simulated Q factor at different VDD

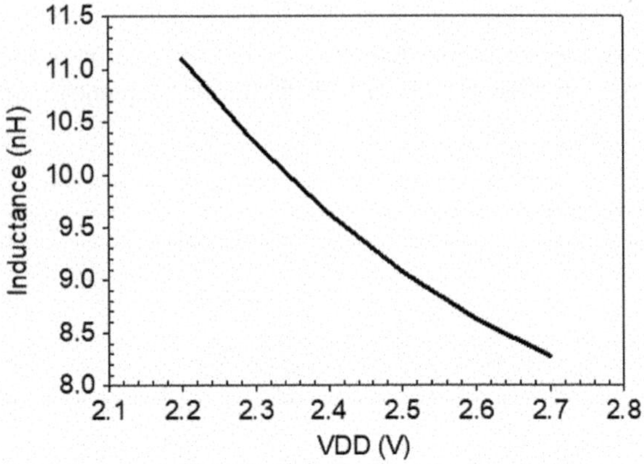

Fig. 6 Variation of inductance with respect to VDD

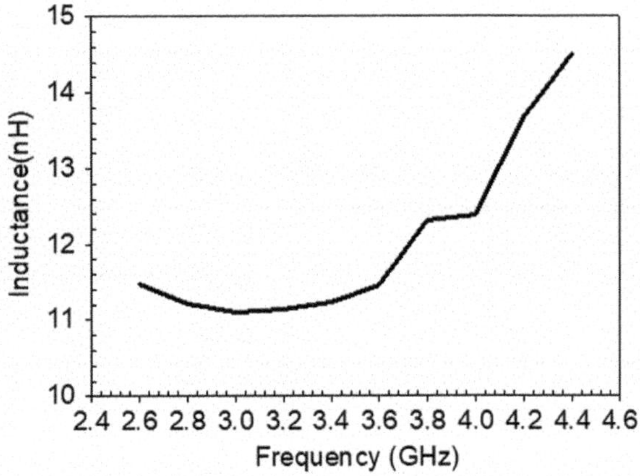

Fig. 7 Variation of inductance with respect to frequency

Table 2 Performance summary of the proposed active inductor with previously published works

Reference	[5]	[14]	[15]	[16]	[17]	This work
Technology (nm)	180	180	180	65	180	180
Supply voltage (V)	1.8	2.0	1.8	1.2	1.9	2.2
Q factor	2.87 k	68 k	21 k	4.81 k	1.40 k	12,580 k
Inductance (nH)	18.7	1.1	5.46	0.25–1.2	14	8.2–14.5
Power consumption (mW)	1.26	5	3.37	2	6	–
Freq range (GHz)	1–10	–	0.9–2.75	20–30	0.1–5.5	2.6–4.4

The performance summary of the proposed DAI with previously released works is displayed in Table 2. It is evident that the proposed DAI outperforms other comparable efforts in terms of Q. However, because of the increased supply voltage of 2.2 V, the power usage is marginally higher. The other performances, however, are interchangeable with other pieces.

The proposed circuit in [5] is implemented by taking advantage of the transistor's intrinsic capacitance and employing a cross-coupled pair of transistors that provide negative feedback for an increased quality factor. The Advanced Design System (ADS) is used to simulate the suggested AI, and a low Q factor of 2.87 k is attained by using an ideal current source in a differential structure. In contrast, the AI in [14] is made up of a single transistor and a passive compensation network, the latter of which permits adjustment of the series resistance and inductance values in order to reach a high Q factor of 68 k. In [15], a cascode double feedback AI based on resistive feedback is proposed. At a frequency of 1.3 GHz, the proposed AI displays an inductance of 5.46 nH and a Q factor of 21 k. In [16], a 65 nm CMOS technique is used to construct an AI with a 1.2 V supply voltage. A Q factor of 4.81 k, an inductance of 0.25 nH to 1.2 nH, and a frequency range of 20 to 30 GHz are attained. Finally, a complementary common-source stage with biasing current and common-mode bias voltage in the differential structure is proposed in [17]. The lowest Q factor of 1.4 k is obtained among the published works.

4 Conclusion

This work employs 018-μm technology to design an extremely high-quality factor completely differential CMOS active inductor with cascoded current mirror for RF front-end applications. The fully differential active inductor with a cascoded current mirror as its current source can achieve the highest Q factor of 12,580 k at 3.6 GHz with an inductance tuning range of 8.2–14.5 nH. Moreover, the Q factor can be tuned by tuning the resistor in the current mirror and supply voltage VDD. In addition, the frequency ranges can be tuned by tuning the resistor value at the differential structure active inductor. The results show that the proposed fully differential active inductor achieves a high Q factor as compared with previously published works and is suitable for RF front-end applications.

Acknowledgements The author would like to acknowledge the support from the Fundamental Research Grant Scheme (FRGS) under a grant number of FRGS/1/2020/TK0/UNIMAP/02/26 from the Ministry of Higher Education Malaysia.

References

1. Niranjan V, Ashwni K, Jain SB (2013) CMOS active inductor for low voltage and low power wireless applications. In: 10th IEEE international conference on wireless and optical communications networks (WOCN) on proceedings, IEEE, India, pp 1–4
2. Murad SAZ, Pokharel RK, Kanaya H, Yoshida K, Nizhnik O (2010) A 2.4-GHz 0.18-μm CMOS Class E single-ended switching power amplifier with a self-biased cascode. AEU-Int J Electron Commun 64(9):813–818
3. Yi KS, Murad SAZ, Mohyar SN (2021) Design of 2.4 GHz two stages cascode class E power amplifier for wireless application. J Phys: Conf Ser 1962:012017. IOP Publishing
4. Meharde P, Niranjan V, Kumar A (2014) Low voltage CMOS active inductor with bandwidth and linearity improvement. In: 5th IEEE international symposium on electronic system design on proceedings, IEEE, India, pp 79–83
5. Noferesti F, Zarei H, Ebrahimi Pour MJ, Bijari A (2019) A fully differential cmos active inductor with high quality factor and high tunability. In: 27th Iranian conference on electrical engineering (ICEE), IEEE, Iran, pp 355–359
6. Ler C, bin A'ain AK, Kordesch AV (2008) Compact, high-*Q*, and low-current dissipation CMOS differential active inductor. IEEE Microw Wirel Compon Lett 18(10):683–685
7. Mariem J, Nabih PEQ (2019) Design of active inductor-based butterworth and chebyshev microwave bandpass filters in standard 0.18μm-CMOS technology. Int J Reconfigurable Embed Syst 8(1):27
8. Kumar R, Sachan D, Yadav SS, Sihara AK, Misra PK (2017) Design of active inductor at 2.4 GHz frequency using 180 nm CMOS technology. In: 4th IEEE uttar pradesh section international conference on electrical, computer and electronics (UPCON) on proceedings, IEEE, India, pp 477–481
9. Slimane A, TEDJINI SA, Haddad F (2018) Novel CMOS active inductor for tunable RF circuits. In: 61st international Midwest symposium on circuits and systems (MWSCAS) on proceedings, IEEE, Canada, pp 917–920
10. Hammadi AB, Mhiri M, Haddad F, Saad S, Besbes K (2017) Study of wide adjustable active inductor circuits: design approaches and reconfiguration methods. In: International conference on engineering & MIS (ICEMIS) on proceedings. IEEE, Tunisia, pp 1–6
11. Saad S, Haddad F, Hammadi AB (2022) High-performance CMOS tunable differential active inductor for RF applications. In: International conference on design & test of integrated micro & nano-systems (DTS) on proceedings, IEEE, Egypt, pp 01–06
12. Singh M, Sarma R (2018) Design and implementation of MOSFET based folded cascode current mirror. In: 2018 international conference on intelligent circuits and systems (ICICS) on proceedings. IEEE, India, pp 17–21
13. Momen HG, Yazgi M, Kopru R, Saatlo AN (2016) A new high performance CMOS active inductor. In: 39th international conference on telecommunications and signal processing (TSP) on proceedings, IEEE, Austria, pp 291–294
14. Leuzzi G, Stornelli V, Pantoli L, Del Re S (2015) Single transistor high linearity and wide dynamic range active inductor. Int J Circuit Theory Appl 43(3):277–285

15. Prameela B, Daniel AE (2020) A novel high Q active inductor design for wireless applications. Procedia Comput Sci 171:2626–2634
16. Delwar TS, Siddique A, Biswal MR, Behera P, Rashed ANZ, Choi Y, Ryu JY (2022) A novel dual mode configurable and tunable high-gain, high-efficient CMOS power amplifier for 5G applications. Integration 83:77–87
17. Nediyara Suresh L, Manickam B (2021) Design and application of CMOS active inductor in bandpass filter and VCO for reconfigurable RF front-end. Integration 82:115–126

Multi-scale Specific Emitter Identification Via Self-attention-Based Feature Pyramid Network

Yurui Zhao, Xiang Wang, and Zhitao Huang

Abstract Specific emitter identification (SEI) authenticates devices via identifying hardware characteristics. To improve identification accuracy and robustness, we design a novel neural network called a self-attention-based feature pyramid network (SA-FPN). The SA-FPN extracts multi-scale fingerprint features via the feature pyramid network (FPN) and makes predictions on the basis of those different scale features. Then, a squeeze and excitation network (SENet), embedded in SA-FPN, fuses predictions with adaptive weights and predicts emitter identities. Experimental results validate that the proposed method achieves higher accuracy and exhibits stronger robustness than other SEI methods.

Keywords Feature Pyramid Network (FPN) · Fingerprint features · Self-attention (SA) · Specific Emitter Identification (SEI) · Squeeze and Excitation Network (SENet)

1 Introduction

Specific emitter identification (SEI) is used to extract fingerprint features via analyzing external characteristics from intercepted signals. Currently, SEI is widely used in many fields, including target tracking [1], spectrum sensing [2], Internet of Things (IoT) security [3], and 6G wireless communication.

With the flourishing of deep learning, many scholars attempt to adopt deep learning in extracting fingerprint features. Depending on neural network inputs,

Y. Zhao · X. Wang (✉) · Z. Huang
College of Electronic Science and Technology, National University of Defense Technology, Changsha, China
e-mail: zhaoyurui@nudt.edu.cn; huangzhitao@nudt.edu.cn

Y. Zhao
e-mail: zhaoyurui@nudt.edu.cn

Z. Huang
e-mail: huangzhitao@nudt.edu.cn

M. Ma (ed.), *Proceedings of the 12th International Conference on Communications, Circuits, and Systems*, Lecture Notes in Electrical Engineering 1193,
https://doi.org/10.1007/978-981-97-2636-3_14

existing methods fall into two major categories. The first one is to feed raw in phase and quadrature (IQ) sequences into a neural network. For extracting fingerprint features from Zigbee devices, Merchant designed a special one-dimension (1D) convolutional neural network (CNN) [4]. Considering stability and accuracy, Weng proposed to modify a residual network (ResNet) into a 1D network [5]. Results indicate that the proposed method can accomplish the SEI task in ADS-B signals successfully. However, when selecting ResNet as feature extractors, dataset augmentation is essential to improve noise robustness. The second one is to feed transforms into neural networks, such as frequency spectrum [6, 7], time-frequency spectrum [8], and constellation diagrams [9]. Aiming at exploiting differences between transforms, Baldini compared recurrent plots (RP), continuous wavelet transforms (CWT), and short-time Fourier transforms (STFT) as the inputs of neural networks [10]. Results argue that wavelet-based approaches outperform other RF-to-image approaches.

However, these neural network-based methods are all data-driven, which makes it difficult to achieve high accuracy in other datasets. In addition, those methods perform less effectively in terms of robustness.

To address the aforementioned issues, we design a novel neural network called a self-attention-based feature pyramid network (SA-FPN) to fuse multi-scale features. The proposed network utilizes the feature pyramid network (FPN) [11] to extract multi-scale features, which describe different hardware characteristics. Based on different scale features, prediction vectors are generated via convolutional layers. Then, a squeeze and excitation network (SENet) [12] fuses prediction vectors and makes a final prediction. To the best of our knowledge, this is the first time that multi-scale neural network features are fused for SEI. Experiments on universal software radio peripherals (USRP) validate that the proposed method can achieve higher accuracy and stronger robustness than other SEI methods.

2 Problem Statement

Ideal communication signals are comprised of three fundamental intentional modulation types, i.e., amplitude modulation, frequency modulation, and phase modulation. Hence, the ideal signal $s_{IM}(n)$ can be written as

$$s_{IM}(n) = A_n \exp\left[j2\pi(f_c + f_n)n + j(\varphi_0 + \varphi_n)\right], \tag{1}$$

where f_c is the carrier frequency and φ_0 is the initial phase. A_n, f_n, and φ_n are respectively the intentional modulation of amplitude, frequency, and phase.

Due to the limitation of production technology, hardware imperfections bring parameter vibrations into the intercepted signal [14], expressed as

$$s_{UIM}(n) = (A_n + \Delta A_n)\exp\begin{bmatrix} j2\pi(f_c + f_n + \Delta f_n)n \\ +j(\varphi_0 + \varphi_n + \Delta\varphi_n) \end{bmatrix}, \tag{2}$$

where ΔA_n, Δf_n, and $\Delta\varphi_n$ are the unintentional modulation of amplitude, frequency, and phase. Besides, wireless communication signals suffer channel noise interruptions when transmitting. And then the intercepted signal $s(n)$ can be expressed as

$$s(n) = (A_n + \Delta A_n)\exp\begin{bmatrix} j2\pi(f_c + f_n + \Delta f_n)n \\ +j(\varphi_0 + \varphi_n + \Delta\varphi_n) \end{bmatrix} + \upsilon_n, \tag{3}$$

where υ_n represents the additive channel noise.

Those three unintentional modulation parameters ΔA_n, Δf_n, and $\Delta\varphi_n$, determined by hardware imperfections, are unique for each emitter. Hence, it is possible to recognize emitter identifies via analyzing parameter characteristics. The SEI process is to extract fingerprint features from the intercepted signals, which imply unique characteristics of unintentional modulation parameters, written as

$$\xi[s(n)] = \zeta(\Delta A_n, \Delta f_n, \Delta\varphi_n), \tag{4}$$

where $\xi(\cdot)$ is the fingerprint feature extraction function and act as the core function in the SEI process. The fingerprint features $\zeta(\Delta A_n, \Delta f_n, \Delta\varphi_n)$ are closely related to hardware imperfections and unique for each emitter. For improving the ability of generalization and robustness, we propose to apply self-attention-based FPN to designing the fingerprint feature extraction function $\xi(\cdot)$.

3 Methodology

The proposed novel network utilizes a ResNet as its backbone network and extracts multi-scale features as identity marks via the FPN. Then, predictions, made on multi-scale features, are fused with a self-attention module and the SENet. The overall architecture is shown in Fig. 1.

As Fig. 1 illustrates, the proposed SA-FPN consists of three main parts, i.e., backbone networks, multi-scale feature extraction, and prediction fusion.

3.1 Backbone Network

The residual neural networks (ResNet) [13], aiming at solving the degradation problem of neural networks, can extract fingerprint features. In the ResNet, the residual component is composed of two convolution layers and an identity mapping, as shown in Fig. 2.

The input and output dimensions of the residual component are the same and can be added directly. There are two reasons to choose ResNet. One is that a residual module can guarantee a low computation complexity with connection skips. The other reason

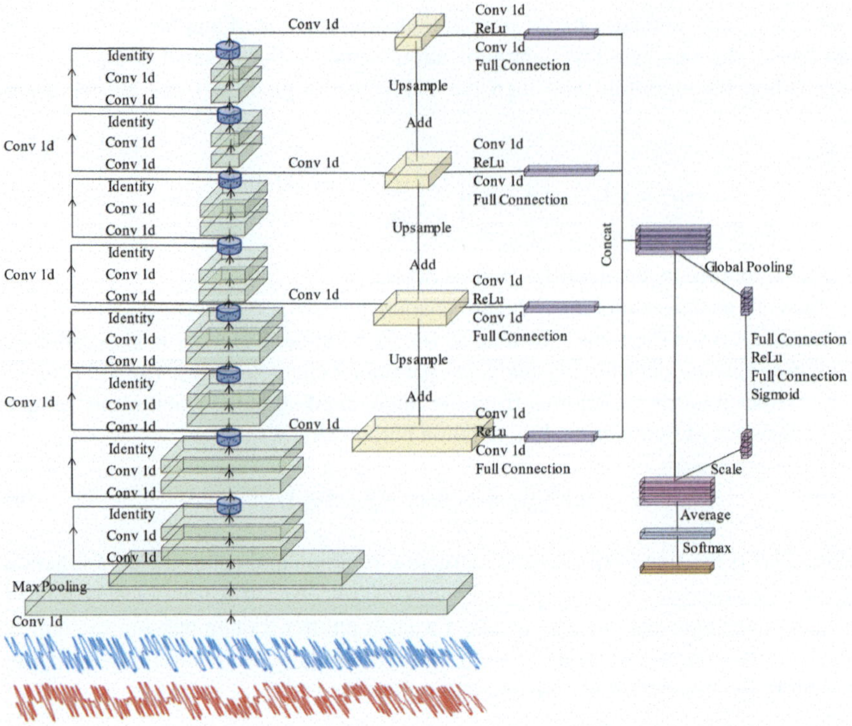

Fig. 1 The architecture of the SEI method based on self-attention-based feature pyramid network (SA-FPN)

is that the same dimension of inputs and outputs makes it easier to fuse multi-scale features. Hence, we choose to use ResNet-18 layers in the present work. Furthermore, to input raw IQ sampling points, we replace the traditional two-dimension (2D) convolution networks with one-dimension (1D) convolution networks.

3.2 Multi-scale Feature Extraction

To extract multi-scale features, we propose to use FPN structures to fuse multi-scale feature maps and extract fingerprint features from different scales, as shown in Fig. 3. An FPN has a top-down architecture with lateral connections and has been developed for building high-level semantic feature maps at all scales. The low-level features are high resolution, while the high-level features are low resolution.

For different scales, the proportion of low-level features and high-level semantic features are different. Hence, we make predictions of device identities via different scale features.

Fig. 2 The structure of a residual component in a residual network (ResNet)

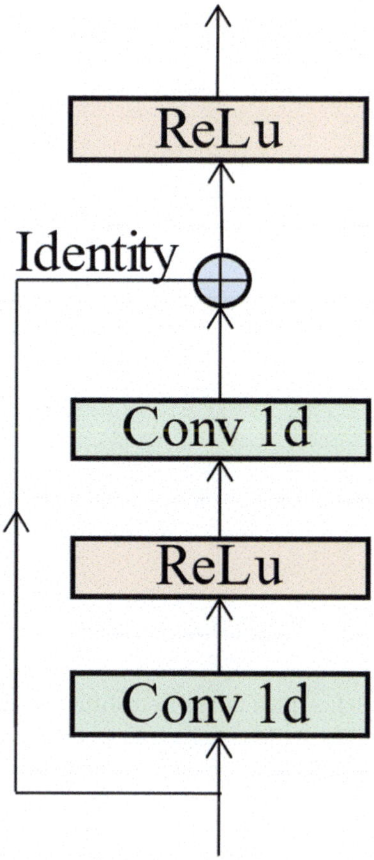

Fig. 3 The structure of a feature pyramid network (FPN)

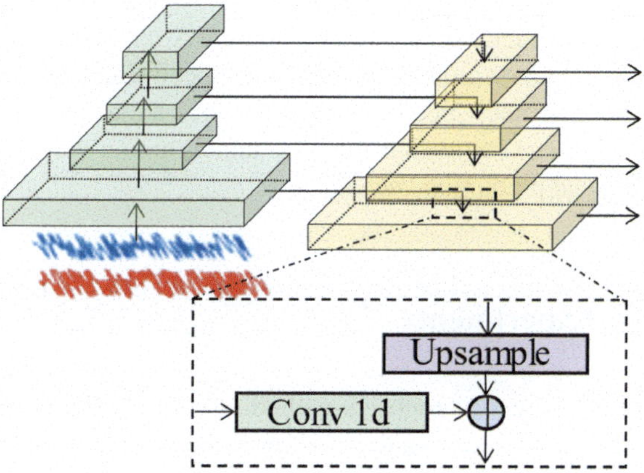

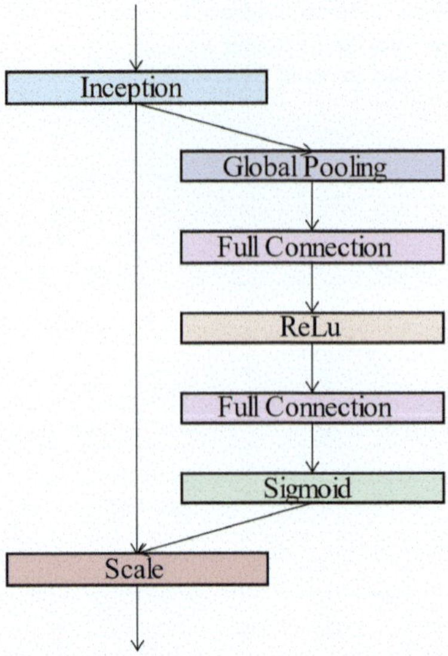

Fig. 4 The structure of a squeeze and excitation network (SENet)

3.3 Prediction Fusion

To fuse predictions from different-scale features, we use an SENet to generate coefficients for linear algebraic combinations. The SENet adaptively recalibrates channel-wise feature responses by explicitly modeling interdependencies between channels, as shown in Fig. 4.

Prediction vectors, generated from different scale features, constitute a feature via concatenation in the channel dimension. Then, the feature is passed through a squeeze operation, which aggregates the feature map across spatial dimensions and calculates channel-wise feature responses. After that, based on channel dependence, an excitation operation learns channel-specific activations by a self-gating mechanism. Those channel-specific activations govern the excitation of each channel. Lastly, the prediction vectors in different channels are re-weighted with activations to generate the final prediction vector.

4 Experiments

To evaluate the proposed method, we collect signals from 15 USRPs and select the Signal Spectrum Analyzer from Rohde and Schwarz as the receiver. For avoiding interruption from outside environments, the USRP and the receiver are directly connected via transmission lines in this experiment.

Each USRP has two independent channels, as shown in Fig. 5a. Hence, the experiment comprises 30 individuals.

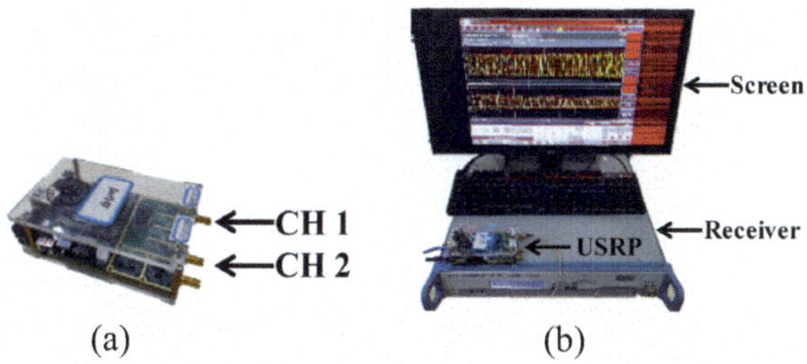

Fig. 5 The experiment environment. **a** USRP plots. Each USRP owns two channels. **b** Signal collection. The receiver is equipped with a screen

In the real-world experiment, the carrier frequency is 770 MHz, the sampling frequency is 19.2 MHz. To test the robustness of channel noise, modulation type, and bit rate, we added additive Gaussian Noise (AWGN) into signals with various modulation and bit rates. The modulation types contain QPSK, 8PSK, 16QAM, and 32QAM, while the bit rates contain 0.96, 1.2, and 1.5 Mbps. The signal-to-noise (SNR) of the received signals varies from 30 dB to 0 dB with a step of 2 dB. In the collection process, we connect the transmitters and receivers with high-frequency coils, where the channel noise can be neglected, as shown in Fig. 5b. For each individual, we collect 5,000 samples, and each sample contains 1,024 raw IQ sampling points.

In order to highlight advantages brought by the SA-FPN structure, we compare the improved architecture together with ResNet [5] when the 1.5 Mbps QPSK signal was 30 dB. Confusion matrixes for these two methods are shown in Fig. 6.

The intensity of the color of the symbols in the figure reflects how the system possibly confuses two classes. As shown in Fig. 6, the proposed method can mitigate mislabeling one emitter as another one. Hence, the accuracy is higher than that of the ResNet-based method. This phenomenon reveals that the fusion of multi-scale features can contribute to extracting the unique characteristics of hardware imperfections.

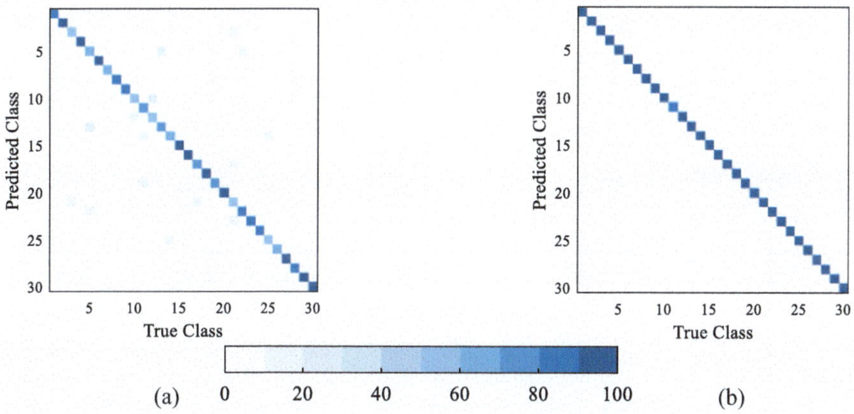

Fig. 6 Confusion matrix with different methods. **a** ResNet. **b** SA-FPN

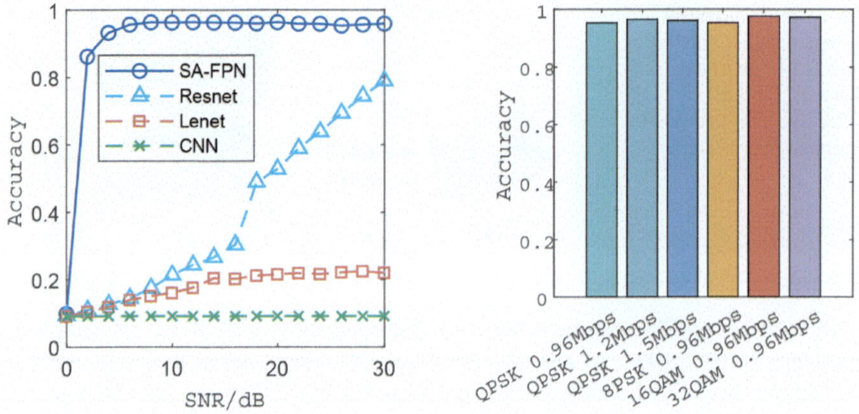

(a) (b)

Fig. 7 Accuracy of the SA-FPN based method. **a** Robustness for noise. **b** Robustness for modulation types and bit rates

For evaluating the robustness for modulation types, bit rates, and channel noises, we perform the proposed method in various scenarios and choose ResNet [5], LeNet [6], and a CNN [4] for comparison. Results are shown in Fig. 7. In Fig. 7a, four methods are performed on the 1.5 Mbps QPSK signals with different noise energies, while in Fig. 7b the proposed method are applied in 20 dB signals with various modulation types or bit rates.

The results plotted in Fig. 7a illustrate that the proposed method outperforms the other methods in accuracy under different SNRs. Hence, compared with recent methods, the proposed SA-FPN can achieve higher accuracy and strengthen the robustness of channel noise. However, the proposed architecture requires more computation consumption due to the parament amount increasement. Also, Fig. 7b reveals that the proposed method can attain proportionable accuracy with various modulation types and bit rates. This phenomenon agrees with the argument that the fusion of different levels can strengthen the network robustness.

5 Conclusion

Aiming at improving identification accuracy and robustness, in this paper, we propose a novel neural network, combining an FPN and SENet, to fuse multi-scale features. Experimental results on USRPs verify that the proposed method can indeed improve accuracy and robustness. To the best of our knowledge, this is the first trial of fusing multi-scale features using neural networks in SEI. In fact, this paper only provides a successful trial in CNN. In further work, this innovation can be applied to other typical neural network modules, such as long short-term memory (LSTM), recurrent neural networks (RNNs), and Transformer.

Acknowledgments This search is supported by the National Natural Science Foundation of China, Grant Number 62271494.

References

1. Li M, Jing Z (2018) Multi-target joint detection, tracking and classification based on generalized Bayesian risk using radar and ESM sensors. arXiv:1807.02267
2. Zeng Y, Liang Y-C, Hoang AT, Zhang R (2010) A review on spectrum sensing for cognitive radio: challenges and solutions. EURASIP J Adv Signal Process 1–15
3. Zhao Y, Wang X, Lin Z, Huang Z (2022) Multi-classifier fusion for open-set specific emitter identification. Remote Sens 14(9):2226
4. Merchant K, Revay S, Stantchev G, Nousain B (2018) Deep learning for RF device fingerprinting in cognitive communication networks. IEEE J Sel Top Signal Process 12(1):160–167
5. Weng LTR, Peng JL, He Y, Zhong DD, Peng JH, Mao XY (2021) Specific emitter identification of ADS-B based on deep residual network. Aero Weapon 28(4):24–29
6. Xiao Y, Wei XZ (2020) Specific emitter identification of radar based on one dimensional convolution neural network. J Phys Conf Ser 1550(3):032114
7. Zhu M, Feng Z, Zhou X (2020) A novel data-driven specific emitter identification feature based on machine cognition. Electronics 9(8):1308
8. Bitar N, Muhammad S, Refai HH (2017) Wireless technology identification using deep convolutional neural networks. In Proceedings of PIMRC 2017, Montreal, QC, Canada, pp 1–6
9. Zha X, Li T, Qiu Z, Feng Y (2021) A novel anti-Doppler SEI algorithm based on the vector diagram decomposition. Electron Lett 57(20):785–787
10. Baldini G, Gentile C, Giuliani R, Steri G (2019) Comparison of techniques for radiometric identification based on deep convolutional neural networks. Electron Lett 55(2):90–92
11. Lin T-Y, Dollar P, Girshick R, He K, Hariharan B, Belongie S (2017) Feature pyramid networks for object detection. In: Proceedings of the IEEE conference on computer vision and pattern recognition, pp 2117–2125
12. Hu J, Shen L, Sun G (2018) Squeeze-and-excitation networks. In: Proceedings of the IEEE conference on computer vision and pattern recognition, 2018, pp 7132–7141
13. Liang J (2020) Image classification based on resnet. J Phys: Conf Ser 1634(1): 012110. IOP Publishing
14. Liu ZM (2021) Multi-feature fusion for specific emitter identification via deep ensemble learning. Digit Signal Process 110:102939

An Ultra-Low Power CMOS Subthreshold Voltage Reference with Temperature Coefficient Compensation

Yuxuan Huang, Ruihuang Wu, Bingjun Xiong, Zhipeng Li, Jia Liu, Yun Zou, Jingjing Liu, and Xinghua Sun

Abstract This paper presents a nanowatt voltage reference (VR) with an ultra-low power consumption temperature coefficient compensation. All transistors are biased in the subthreshold region to reduce the power consumption of the proposed VR circuit. The complementary-to-absolute-temperature (CTAT) voltage and proportional-to-absolute-temperature (PTAT) voltage are mainly generated by two NMOS transistors with different threshold voltages. At the same time, a simple temperature compensation circuit is designed to optimize the temperature coefficient at high temperatures, so that the circuit can generate a reference voltage with a wider temperature range and a lower temperature coefficient. The proposed VR circuit is fabricated in a 0.18-μm CMOS process with a silicon area of only 0.022 mm^2. Theoretical analysis and post-layout simulation results verified the reliability and performance improvement of circuit operation. The proposed VR circuit generates a

Y. Huang · R. Wu · B. Xiong · Z. Li · J. Liu · Y. Zou · J. Liu (✉) · X. Sun
School of Electronics and Communication Engineering, Sun Yat-Sen University, Shenzhen, China
e-mail: liujj77@mail.sysu.edu.cn

Y. Huang
e-mail: huangyx295@mail2.sysu.edu.cn

R. Wu
e-mail: wurh29@mail2.sysu.edu.cn

B. Xiong
e-mail: xiongbj@mail2.sysu.edu.cn

Z. Li
e-mail: lizhp57@mail2.sysu.edu.cn

J. Liu
e-mail: liujia49@mail.sysu.edu.cn

Y. Zou
e-mail: zouy83@mail2.sysu.edu.cn

X. Sun
e-mail: sunxinghua@mail.sysu.edu.cn

M. Ma (ed.), *Proceedings of the 12th International Conference on Communications, Circuits, and Systems*, Lecture Notes in Electrical Engineering 1193,
https://doi.org/10.1007/978-981-97-2636-3_15

reference voltage of 308.21 mV with a temperature coefficient of 6.42 ppm/°C over a wide temperature range of −40 ~ 40 °C. The power consumption of the circuit is 7.04 nW. The voltage line sensitivity (LS) is 0.099%/V.

Keywords Voltage reference · Low power · Low supply voltage · Temperature compensation · Temperature coefficient · Subthreshold

1 Introduction

Nowadays, the Internet of Things (IoT) technology is developing rapidly. The IoT sensors are developed for small size, low voltage, and low power, or even self-powered by harvesting energy from the environment. So, we need to design a voltage reference circuit with good performance and meet the requirements of low voltage and low power consumption. A voltage reference circuit is a very basic module in the analog circuit field. The performance of a voltage reference circuit often affects the overall performance of the entire system. The VR circuit needs to generate a stable voltage independent of process, power supply voltage, and temperature under the condition of low and unstable power supply voltage and low power consumption. Therefore, sub-threshold CMOS voltage reference circuits have attracted more and more attention in recent years.

Sub-threshold CMOS circuits using leakage currents are becoming more widely used [1], as well as in reference circuits. In 2012, with the introduction of a two-transistor voltage reference using leakage currents, the power consumption and power supply of the VR circuit were reduced without the use of bipolar junction transistors (BJT) [2]. Then, based on the two-transistor structure, the bias current can be set and a stable reference circuit can be built by taking advantage of the virtual short characteristic of the input end of the operational amplifier (OPA). However, their power consumption is relatively high [3–6]. In 2020, a self-biased reference circuit based on the two-transistor structure without amplifiers and start-up circuits was proposed. It has advantages in power consumption and start-up time, but the temperature coefficient is only 62 ppm/°C [7]. In 2022, a voltage reference circuit combining sub-threshold transistors and BJTs was proposed, but due to the use of BJT transistors, power consumption is relatively high, and minimum power supply voltage is limited [8].

To solve some of the problems mentioned above, this work proposes a CMOS subthreshold voltage reference circuit using a temperature compensation circuit, without BJT transistors. The rest of the paper is organized as follows, Section 2 discusses the proposed design and working mechanism. Section 3 describes the simulation results and the performance of the proposed VR circuit. Section 4 concludes the work.

2 The Proposed Voltage Reference Circuit

The CMOS subthreshold voltage reference circuit proposed in this work is shown in Fig. 1. This circuit includes a start-up circuit, a bias circuit, an OPA circuit, a VR core circuit, and a temperature compensation circuit. The start-up circuit includes M_{S1}, M_{S2}, and M_{S3}, which can make the circuit deviate from the original zero bias point into the normal operation state, wherein M_{S1} acts as a capacitor. When the power supply is just powered on, the voltage reference output voltage value is zero. At this time M_{S3} is turned off, through continuous charging. When the gate voltage of M_{S2} increases, it causes M_{S2} to be turned on. And then all the gate voltages of M_{P5} and M_{P6} are pulled down. The two branches of the VR core circuit start to flow current, and this current is copied to each branch of the voltage reference circuit through the current mirrors. Finally, the circuit starts to work normally. When the output voltage reference is greater than the V_{TH} of M_{S3}, M_{S3} is turned on and the gate voltage of M_{S2} is pulled down. So M_{S2} is turned off and the start-up circuit is turned off. The low-temperature coefficient of the circuit is mainly realized by the temperature compensation circuit, which adjusts the voltage deviation of the output voltage of the voltage reference core circuit at high temperatures.

2.1 *Operating Principle*

All the transistors in the circuit are biased in the sub-threshold region. The purpose is to reduce the power supply voltage and power consumption. Among them, M_{N8} and M_{N11} are thick-oxide NMOS, and the rest are thin-oxide transistors. The main part of the voltage reference circuit can generate PTAT voltage and CTAT voltage for first-order temperature coefficient compensation. The drain current of the subthreshold transistor is as follows:

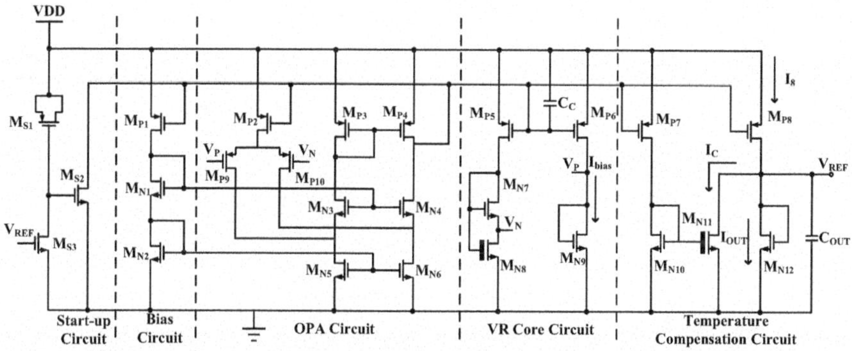

Fig. 1 Proposed CMOS subthreshold voltage reference

$$I_D = \mu_n C_{OX}(m-1)KV_T^2 \exp\left(\frac{V_{GS}-V_{TH}}{mV_T}\right)\left(1-\exp\left(\frac{-V_{DS}}{V_T}\right)\right) \quad (1)$$

where μ_n is the mobility of the electrons, C_{OX} is the gate-oxide capacitance per unit area, V_{TH} is the threshold voltage of the transistor, K is the aspect ratio (=W/L) of the transistor, $V_T = k_B T/q$ is the thermal voltage, k_B is the Boltzmann constant, and T is absolute temperature, q is the elementary charge, and m is the subthreshold slope factor. In the derivation process, it can be approximated that the C_{OX} and V_{TH} of the same type of NMOS transistors are equal, and the subthreshold slope factor m of all NMOS transistors is the same. When $V_{DS} \geq 3V_T$, the drain current of the transistor in the subthreshold region can be approximated as:

$$I_D = \mu_n C_{OX}(m-1)KV_T^2 \exp\left(\frac{V_{GS}-V_{TH}}{mV_T}\right) \quad (2)$$

Since the drain current of M_{N7} and M_{N8} are equal, $I_{DN7} = I_{DN8}$. There is a threshold voltage difference between M_{N7} and M_{N8}, the voltage of node N is as follows:

$$V_N \approx (V_{THN8} - V_{THN7}) + mV_T \ln\left(\frac{C_{OXN7}K_{N7}}{C_{OXN8}K_{N8}}\right) \quad (3)$$

It can be seen that the voltage of the node N is independent of the branch current. Due to the use of NMOS transistors with different gate oxide thicknesses, there is a threshold voltage difference ΔV_{TH} between them, and the difference in threshold voltage generates a CTAT voltage:

$$\Delta V_{TH} = \Delta V_{TH0} + (\alpha_{N8} - \alpha_{N7})(T - T_0) \quad (4)$$

where α is the first-order temperature coefficient of the threshold voltage, and $\alpha < 0$. ΔV_{TH0} is the threshold voltage difference at normal temperature, and T_0 is the normal temperature of 300 K. At the same time, the ΔV_{TH} has a negative temperature coefficient reduced by $(\alpha_{N8} - \alpha_{N7})$. The slope of the thermal voltage V_T generating PTAT voltage does not need to be too large to offset the first-order temperature coefficient. So, it can also reduce the number of PTAT stages. According to the characteristics of the operational amplifier, the voltage at node P is forced to be the same as the node N, so the bias current I_{bias} is:

$$I\text{bias} = \mu_n \frac{C_{OXN9}C_{OXN7}}{C_{OXN8}}(m-1)\left(\frac{K_{N9}K_{N7}}{K_{N8}}\right) V_T^2 \exp\left(\frac{V_{THN8} - V_{THN7} - V_{THN9}}{mV_T}\right) \quad (5)$$

The final output voltage reference is generated by I_{OUT}, and $I_{OUT} = I_8 - I_C$, where I_8 is the current copied by I_{bias} through the current mirror. I_C is the compensation current generated by the temperature compensation circuit, which is used to compensate for the temperature coefficient at high temperatures. Finally, the voltage reference output value generated by I_{OUT} through the active load M_{N12} is as follows:

$$V_{REF} \approx V_{THN12} + mV_T \ln\left[\frac{C_{OXN7}K_{N7}K_{N9}K_{P8}}{C_{OXN8}K_{N8}K_{P6}K_{N12}} \exp\left(\frac{V_{THN8} - V_{THN7} - V_{THN9}}{mV_T}\right) - \frac{K_{N7}K_{N9}K_{P7}K_{N11}}{K_{N8}K_{P6}K_{N10}K_{N12}} \exp\left(\frac{-V_{THN7}}{mV_T}\right)\right] \quad (6)$$

It can be seen that in the low-temperature range, V_T is small, and exp(-V_{THN7}/mV_T) = K_C can be ignored. Then (6) can be approximate to be (7).

$$V_{REF} \approx V_{THN8} - V_{THN7} + mV_T \ln\left[\frac{C_{OXN7}K_{N7}K_{N9}K_{P8}}{C_{OXN8}K_{N8}K_{P6}K_{N12}}\right] \quad (7)$$

In the low-temperature range, V_{REF} is similar to the voltage of node N. The CTAT voltage and the coefficient-controllable PTAT voltage can be superimposed to eliminate the first-order temperature coefficient by adjusting the transistors' size. When the circuit is in a high-temperature range, the voltage is no longer dominated by the first-order temperature coefficient. As the temperature rises, the temperature coefficient needs to be compensated. With the continuous increase of V_T, K_C can no longer be ignored and the continuously increasing voltage reference output is reduced. Therefore, the temperature performance and operating temperature range of the circuit are effectively improved. The PTAT voltage, which is used to compensate for the first-order coefficient of CTAT, depends on K_{N7}, K_{N8}, K_{N9}, K_{P6}, and K_{P8} parameters. The current I_C is used to compensate for the high-temperature coefficient of the circuit, which is determined by K_{N7}, K_{N8}, K_{N9}, K_{P6}, K_{P7}, K_{N10}, K_{N11}, and K_{N12} parameters. At the same time, the power consumption of the entire circuit is also determined by each transistor's size, so the optimal size of each transistor is a balance between performance and power.

2.2 Analysis of LS

Both LS and PSRR are key indicators that need to be paid attention to in voltage reference circuits. In low-voltage and low-power applications, the operating bandwidth of transistors in the subthreshold region is low, resulting in poor PSRR performance and greater impact from power supply voltage disturbances. Therefore, the improvement of these indicators is particularly critical. The index LS to evaluate the dependence of V_{REF} on the change of direct current (DC) voltage is defined as:

$$LS = \frac{\Delta V_{REF}}{\Delta VDD \times V_{REF}} \times 100\% \quad (8)$$

where ΔVDD and ΔV_{REF} are respectively the power supply voltage range in which the reference voltage circuit works normally and the corresponding reference voltage output variation range. It can be seen from the derivation of (6) that as long as the drain-source voltage V_{DS} of the NMOS transistor in the circuit is greater than or equal to $3V_T$, the reference voltage output does not depend on the change of supply voltage. However, in the derivation process, the current mirror circuits are regarded as ideal devices, and all actual LS will depend on the LS of the current mirror transistors.

3 Post-simulation Results

The proposed voltage reference circuit is implemented by all MOSFET, and designed using a standard 0.18-μm CMOS process. Figure 2 shows the circuit layout, the area occupied by the circuit is 0.022 mm^2, which includes 4.5 pF output capacitance C_{OUT} and 1.7 pF compensation capacitance C_C. Figure 3 shows the simulation results of the temperature performance with and without the temperature compensation circuit. It can be seen that without the temperature compensation circuit, V_{REF} rises rapidly at high temperatures, limiting the circuit's temperature performance and operating temperature range. The proposed VR circuit has an improved temperature coefficient in the high-temperature range. The temperature compensation circuit compensates for the rapidly increasing V_{REF} under high-temperature conditions and realizes the temperature coefficient compensation. Finally, the temperature coefficient of the circuit is reduced to 6.42 ppm/°C, and the operating temperature range of the proposed VR circuit is increased to −40 ~ 140 °C.

Figure 4 shows the post-simulation results of voltage line sensitivity. When the power supply voltage is 0.7–2.2 V, the LS of the proposed VR circuit is only 0.099%/V at room temperature (25 °C). Figure 5 shows the power consumption performance of the mentioned voltage reference circuit at different temperatures when the power supply voltage is 1 V. It can be seen that the power consumption at room temperature is as low as 7.04 nW. Figures 6 and 7 show the start-up time at 25 °C and −40 °C, respectively. The start-up time of the circuit is about 1.3 ms at room temperature.

Table 1 summarizes the performance comparison of low-power reference circuits that have been reported in recent years. With the introduction of an efficient temperature compensation circuit, the circuit can achieve a good temperature coefficient performance in a wide temperature range of −40 °C to 140 °C, while also maintaining low power consumption. Additionally, as the proposed design utilizes an OPA, the circuit exhibits low LS. After integrating the start-up circuit, the start-up time at room temperature has been reduced to 1.3 ms. At the same time, even if the circuit adds a compensation capacitor and an output capacitor, it occupies a smaller area in the same type of circuit.

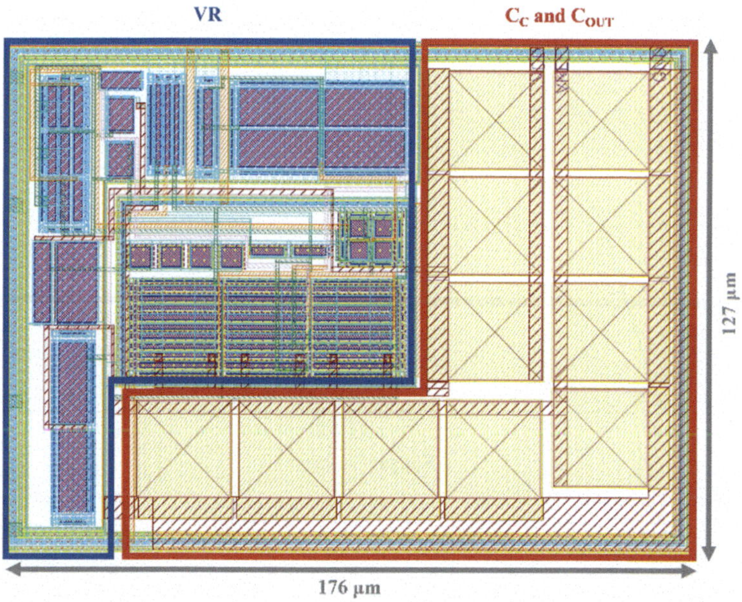

Fig. 2 Layout of the proposed reference

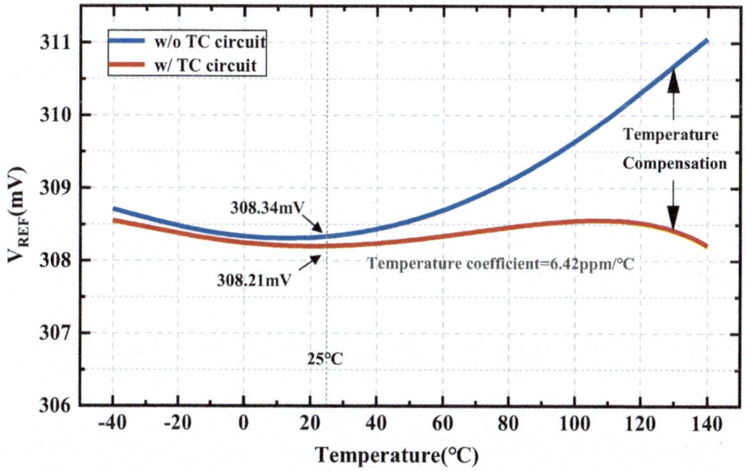

Fig. 3 Comparison of simulated V_{REF} with and without temperature compensation circuit

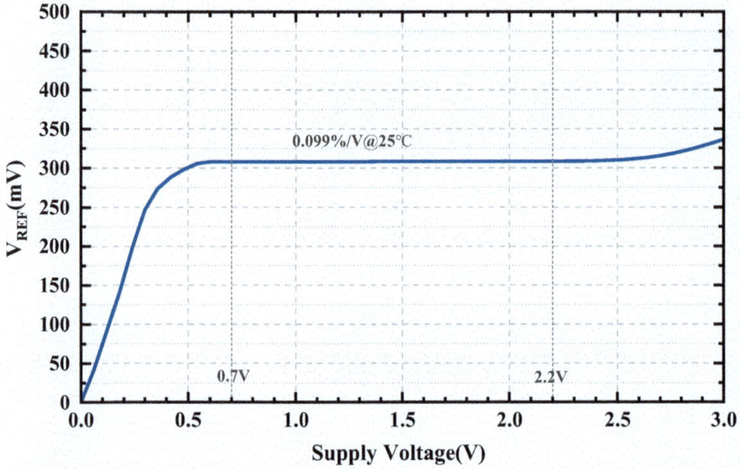

Fig. 4 Post-simulation of LS of the proposed circuit at room temperature

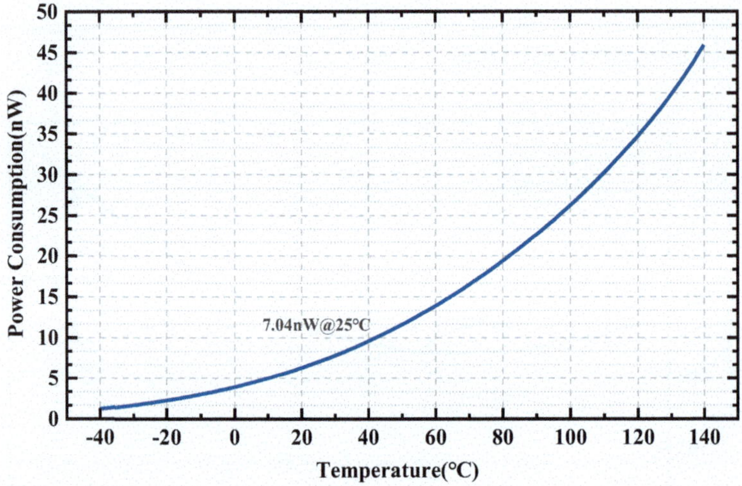

Fig. 5 Post-simulated power consumption at different temperatures

4 Conclusion

This paper presents a subthreshold voltage reference circuit with an efficient temperature coefficient compensation. The voltage reference is generated by two NMOS transistors with different threshold voltages to preliminarily offset of low-order temperature coefficient. A simple and low-power temperature compensation circuit is employed to further enhance its temperature performance and range. The circuit is fabricated in a 0.18-μm CMOS process. A theoretical analysis has been conducted.

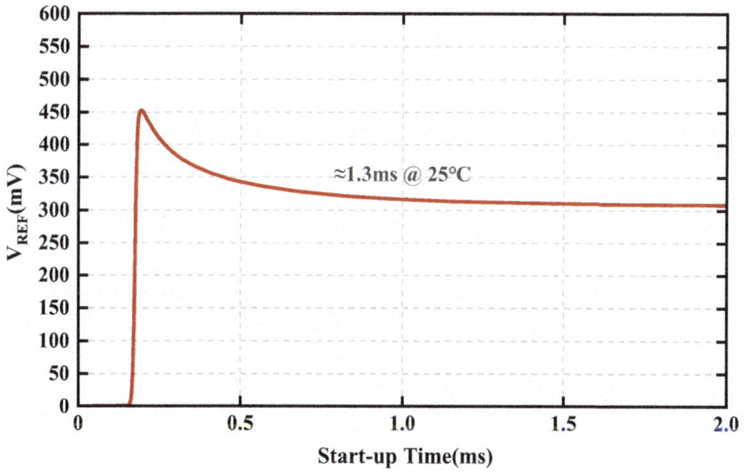

Fig. 6 Post-simulated start-up time at room temperature

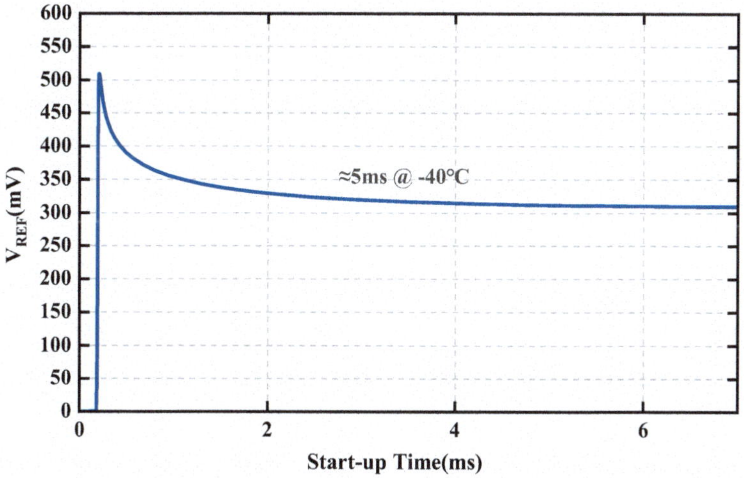

Fig. 7 Post-simulated start-up time at −40 °C

Post-layout simulation results demonstrate that, even under significant variations in power supply voltage and temperature conditions, the circuit maintains a relatively stable voltage output. Additionally, the power consumption of this circuit is very low, which has advantages compared to similar circuits. In conclusion, this circuit offers a practical solution for low-voltage, low-power IoT sensor applications.

Table 1 Performance summary and comparison with other works

	This work [a]	JSSC' 2021 [7][b]	TCASII' 2018 [6][b]	ISCAS' 2022 [8][c]	ISCAS '2020 [9][b]	JSSC' 2013 [10][b]	TCASII'2018 [11][b]
Technology (nm)	180	180	180	130	180	180	180
Type	CMOS	CMOS	CMOS	Sub-BGR	Sub-BGR	Sub-BGR	CMOS
Supply Voltage (V)	0.7–2.2	0.9–1.8	0.4–1.8	0.9	0.65	0.7–1.8	0.6–2.0
Power (nW)	7.04	1.8	9.6	30	1	52.5	30.5
V_{REF}(mV)	308.21	261	210	474	260	548	218.3
Temp. Range (°C)	–40 ~ 140	–40 ~ 130	−40 ~ 140	−20 ~ 80	−50 ~ 85	–40 ~ 120	−40 ~ 125
TC (ppm/°C)	6.42	62	82	19	95	114	23.5
LS (%/V)	0.099	0.013	0.027	0.1	0.23	N/A	0.4
1% Settling Time (ms)	1.3	0.2	N/A	N/A	N/A	6	N/A
Area (mm^2)	0.022	0.005	0.021	0.04	0.075	0.0246	0.075

[a] Post-simulation results. [b] Measured results. [c] Simulation results

Acknowledgements This work is supported by the National Science Foundation of China with project number 62174181.

References

1. Anjana R, Somkuwar AK (2012) Minimizing the sub threshold leakage for high performance CMOS circuits using novel stacked sleep transistor technique. Int J Electr Electron Eng Telecommun 1(1):76–82
2. Seok M, Kim G, Blaauw D, Sylvester D (2012) A portable 2-transistor picowatt temperature-compensated voltage reference operating at 0.5 V. IEEE J Solid-State Circuits 47(10):2534–2545
3. Wang L, Zhan C (2019) A 0.7-V 28-nW CMOS subthreshold voltage and current reference in one simple circuit. IEEE Trans Circuits Syst I: Regul Pap 66(9): 3457–3466
4. Wang L, Zhan C, Lin J, Zhao S, Zhang N (2021) A 0.9-V 22.7-ppm/°C sub-bandgap voltage reference with single BJT and two resistors. In: 2021 IEEE international symposium on circuits and systems (ISCAS), Daegu, Korea, 2021, pp 1–4
5. Wang L, Zhan C, Tang J, Liu Y, Li G (2018) A 0.9-V 33.7-ppm/°C 85-nW sub-bandgap voltage reference consisting of subthreshold MOSFETs and single BJT. IEEE Trans Very Large Scale Integr (VLSI) Syst 26(10):2190–2194
6. Liu Y, Zhan C, Wang L, Tang J, Wang G (2018) A 0.4-V wide temperature range All-MOSFET subthreshold voltage reference with 0.027%/V line sensitivity. IEEE Trans Circuits Syst II: Express Briefs 65(8):969–973
7. Shao C-Z, Kuo S-C, Liao Y-T (2021) A 1.8-nW, −73.5-dB PSRR, 0.2-ms startup time, CMOS voltage reference with self-biased feedback and capacitively coupled schemes. IEEE J Solid-State Circuits 56(6):1795–1804
8. Shetty D, Steffan C, Bösch W, Grosinger J (2022) Ultra-low-power IoT 30nW 474mV 19 ppm/°C voltage reference and 2 nA 470 ppm/°C current reference. In: 2022 IEEE international symposium on circuits and systems (ISCAS), Austin, TX, USA, 2022, pp 843–847
9. Chowdary G, Kota K, Chatterjee S (2020) A 1-nW 95-ppm/°C 260-mV startup-less bandgap-based voltage reference. In: 2020 IEEE international symposium on circuits and systems (ISCAS), Seville, Spain, 2020, pp 1–4
10. Osaki Y, Hirose T, Kuroki N, Numa M (2013) 1.2-V supply, 100-nW, 1.09-V bandgap and 0.7-V supply, 52.5-nW, 0.55-V subbandgap reference circuits for nanowatt CMOS LSIs. IEEE J Solid-State Circuits 48(6):1530–1538
11. Huang C, Zhan C, He L, Wang L, Nan Y (2018) A 0.6-V minimum-supply, 23.5 ppm/°C subthreshold CMOS voltage reference with 0.45% variation coefficient. IEEE Trans Circuits Syst II: Express Briefs 65(10): 1290–1294

Design and Implementation of a Low-Power JESD204B Transmitter with Wideband Fractional-N Ring PLL in 40-Nm CMOS Process

Hao Chen, Jun Deng, Peng Fei Liao, Shan Shan Zhao, Qiushi Wang, and Fang Tang

Abstract This paper presents a transmitter circuit for high-speed data transmission in ADC applications, providing a wide frequency output range of 1–16 Gbps through parallel-to-serial conversion. A source-series-terminated (SST) transmitter achieves 1.2 Vpp output swing and incorporates a voltage regulator structure to mitigate power supply noise interference. The output frequency range is extended through the use of a two-stage differential ring oscillator and a sigma-delta fractional divider. Fabricated in 40-nm CMOS, the transmitter occupies an area of 0.074 mm^2 and achieves an output frequency range of 1–16 Gbps while operating over channels with 13.5 dB loss at Nyquist, consuming 3.8 mW/Gb/s.

Keywords High-speed interface circuit · Fractional PLL · Voltage mode logic (VML) · Current mode logic (CML)

H. Chen · J. Deng · Q. Wang · F. Tang (✉)
School of Microelectronics and Communication Engineering, Chongqing University, Chongqing, China
e-mail: frankfangtang@gmail.com

H. Chen
e-mail: haonan2713@cqu.edu.cn

J. Deng
e-mail: dj_richard@cqu.edu.cn

Q. Wang
e-mail: 370379331@qq.com

P. F. Liao · S. S. Zhao
Chongqing Acoustical-Optical-Electrical Group Co, Electronics Technology Group Corp, Chongqing, China
e-mail: fly2159@163.com

S. S. Zhao
e-mail: zhaoss@bu.edu

M. Ma (ed.), *Proceedings of the 12th International Conference on Communications, Circuits, and Systems*, Lecture Notes in Electrical Engineering 1193,
https://doi.org/10.1007/978-981-97-2636-3_16

1 Introduction

Historically, the high-speed data transmission of ADCs has relied on the use of low-voltage differential signaling (LVDS) to recover the data at the receiving end by synchronizing the clock. However, with the emergence of the JESD204B transmission standard, a new method of high-speed ADC data transmission has been introduced [1]. This method involves packaging the data using the JESD204B protocol and transmitting it to the remote end via a parallel-to-serial conversion process performed by serializer/deserializer (SerDes) circuits, without the need for an additional clock signal. At the receiving end, the clock data restorer is used to recover the synchronous clock, which can address the issue of data synchronization across multiple channels. Additionally, integrating equalizers at the transmitting and receiving ends can mitigate the inter-code crosstalk problem caused by channel fading during channel transmission.

While high-speed data transmission via JESD204B has many advantages, low-speed ADC data transmission still primarily relies on synchronous LVDS transmission with a clock [2, 3], due to its cost-effectiveness and ease of integration. Bridging the gap between these two applications, this paper proposes a transceiver system that uses LVDS parallel transmission and serializes it for high-speed data transmission to address the signal integrity challenges of long-distance transmission, while achieving a low-power and cost-effective design [4].

In this paper, we propose a SST transmitter to achieve serial data transmission at rates of 1–8 Gbps, with support for dual-channel transmission and a total transmission bandwidth of 16 Gbps. Our design leverages a two-stage differential ring oscillator and sigma-delta fractional divider to improve the output frequency range, and employs a voltage regulator structure to mitigate noise interference from the power supply. The transmitter is fabricated using a 40 nm CMOS process and the test results meet the transmission standard requirements of SerDes.

Section 1 of the article introduces the architecture of the JESD204B transmitter; Sect. 2 describes the design methodology of each module in the transmitter, including the SST driver, the digital protocol, and the clock; Experimental results from a 40-nm CMOS prototype are presented in Sect. 4. Finally, Sect. 5 concludes the paper.

2 Transmitter Architecture

2.1 Transmitter Architecture Overview

The proposed transmitter, illustrated in Fig. 1, employs a dual-lane architecture, with each lane featuring an LVDS receiver, JESD204B interface, 40:1 multiplexer, SST driver, PLL, and built-in pseudo-random bit sequence (PRBS) generator. The term "CX" denotes the clock frequency of a block, which corresponds to a clocking speed equal to 1/Xth of the data rate. For instance, "C2" indicates a quarter-rate clock

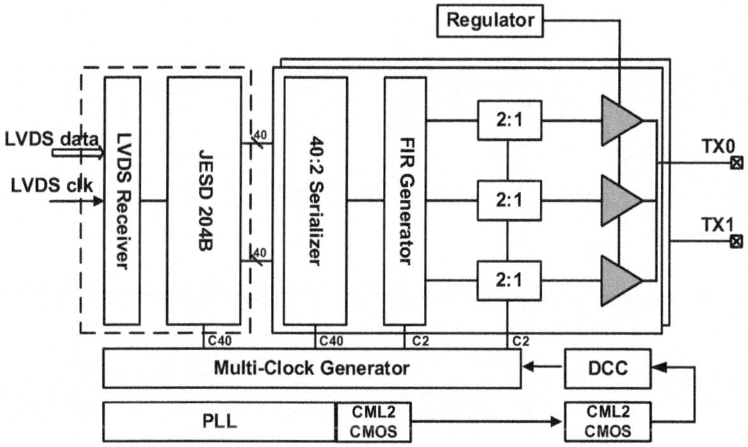

Fig. 1 Block diagram of the proposed transmitter

of 4 GHz. The ADC data is first received in the form of LVDS, which supports a transfer rate of up to 1.6 Gbps. Subsequently, the LVDS receiver converts the incoming signal into a CMOS signal, and the signal is further processed through JESD204B before being directed to the 40:1 multiplexer. Ultimately, the SST driver delivers the output signal. The PLL serves as the clock signal for each module, and its performance, primarily phase noise, governs the timing margin of the digital part and also affects the output signal's jitter size. To address the timing issues associated with the signal transfer between different modules, the current study designs two FIFOs (first-in-first-out), one located between LVDS and JESD204B, and the other between JESD204B and multiplexer. FIFO usage, rather than synchronous data transmission, can effectively resolve the clock and data synchronization issues during long-distance transmission. Nonetheless, it is necessary to consider the potential for FIFO overflow resulting from clock deviation, which may lead to data loss. Accordingly, the proper design of the FIFO depth constitutes a key factor.

2.2 *Digital Protocols*

The JESD204B component is meticulously designed in adherence to the standard protocol [5], comprising an 8b/10b encoder, a scrambler, and an alignment module for multi-channel data synchronization. Moreover, to achieve self-diagnosis capability, a PRBS generator and checker circuit are also developed for this component. In terms of ADC data transmission modes, the multiplexed and interleaved modes are the most commonly utilized [6]. The receivers presented in this paper provide support for multi-mode, multi-channel, and variable-resolution ADC data. As a case in point, the multiplexed mode configuration is presented in Fig. 2, which allows for the support

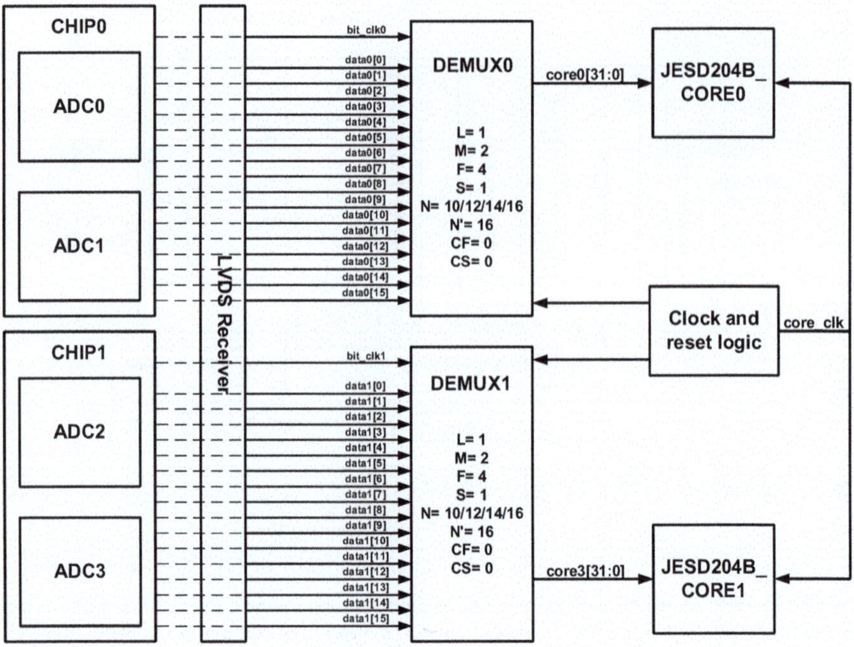

Fig. 2 Block diagram of interference between ADC and JESD204B

of a 4-channel ADC with each 32 LVDS channels representing the sampling data of one ADC. The Demultiplexer of the digital module is tasked with transforming the 16-channel data into 32 channels. Notably, the principal challenge of this module is the inherent skew issue of the LVDS signals across various channels. With up to 32 LVDS input channels to manage, package and case design, PCB layout, and chip layout are executed independently to meet the overall timing requirements.

3 Circuit Implementation

3.1 Transmitter Circuit

As shown in Fig. 3, the transmitter consists of two 20:1 MUX, a 2:1 MUX with a half-rate clock, and a full-rate SST driver with a 3-tap FFE for low power consumption. Considering the data rate limitation, the 2:1 MUX structure is different from the 20:1 MUX which uses CMOS logic, uses a pseudo-differential structure. Without a full-rate clock retiming after 2:1 MUX, systematic and random offsets with PVT variation will lead to Clock duty cycle distortion, which will degrade the transmitter performance. As shown in Fig. 6, to improve the clock performance, a duty cycle correction circuit is applied to the Clock buffer.

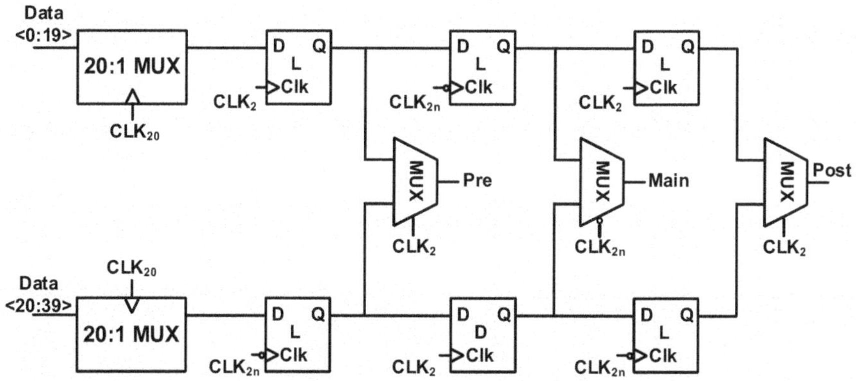

Fig. 3 Three delayed data generator with half-rate 2:1 MUX for 3-tap FFE

The circuit structure of Latch is shown in Fig. 3, which uses the TSPC base unit, and the differential structure is used. The biggest advantage of this is that it increases the noise immunity of the circuit, but it also increases the area of the chip, so it is only used in this paper for the last stage.

In order to isolate the noise from the power supply (which affects the deterministic jitter of the Transmitter output), and also to reduce the influence of switch noise in the driver section on the other module circuits, an on-chip low-dropout regulator (LDO) is used for driver [7, 8].

As shown in Fig. 4, the output driver uses slicer switching to control the output impedance and amplitude. Each slicer is designed as shown in the dashed box of Fig. 4 and can be selected for different path signal inputs. The general standard requires the return loss of the output to be less than −12 dB. We combine the change of resistance with PVT in this process and design 36 slicers to achieve the smallest possible output impedance adjustment accuracy.

However, it is important to note that a larger number of slicers does not necessarily translate to better performance. Rather, the load situation of the driver pre-stage MUX must be taken into consideration. To this end, a shared slicer is utilized by the post-cursor and main-cursor, enabling them to select different paths through the MUX and thereby alleviate the load pressure on the pre-stage.

3.2 *Clock Circuit*

In the SerDes transmitter, the phase-locked loop and the clock transfer circuit are the key modules that affect the output jitter performance. The phase noise of the phase-locked loop output clock is passed directly to the output data, which shows up as jitter in the eye diagram. As shown in Fig. 5, the noise passed to the transmitter is divided into two parts, where the low-frequency noise of $1/f^2$ and $1/f^3$ is mainly contributed

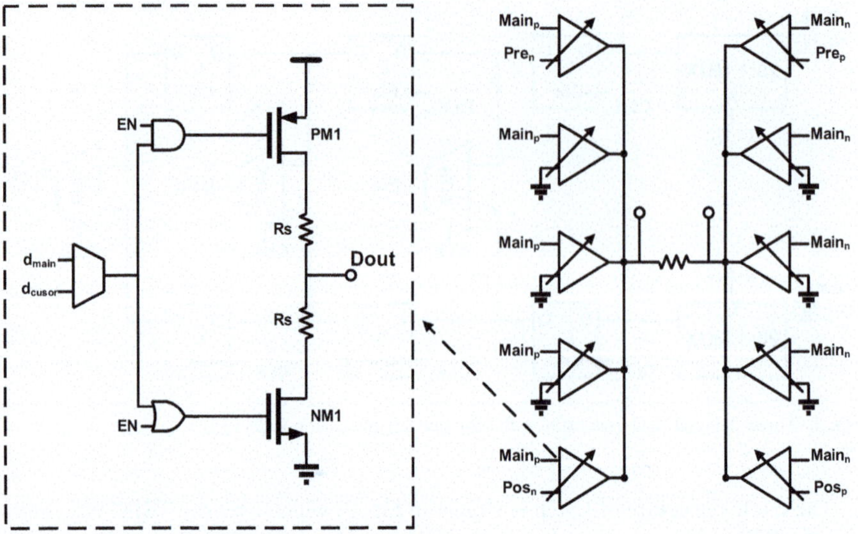

Fig. 4 Block diagram of SST driver with 3-tap FFE

by the PLL [9], while the white noise or bottom noise in the high-frequency part is generated by the clock buffer. In SerDes data transmission, the noise in the low-frequency part is filtered out by the CDR of the receiver, so the noise optimization is mainly focused on the frequency range above the CDR bandwidth.

The noise impact generated by the PLL can be evaluated by integrating the noise over the frequency range, as shown in the formula (1), and the RMS jitter value contributed by the PLL can be obtained based on the noise frequency data fitted to the PLL.

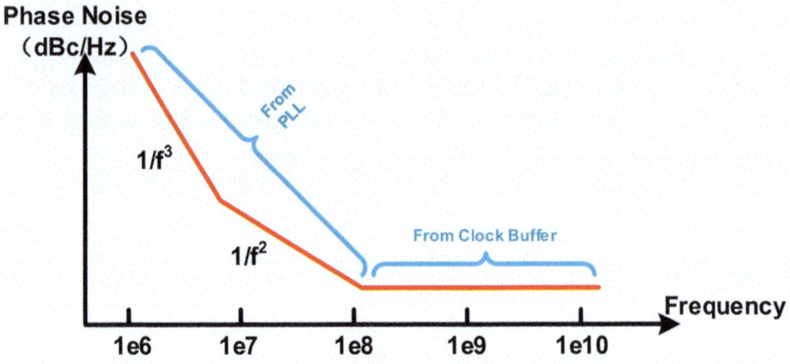

Fig. 5 Clock phase noise at the transmitter output

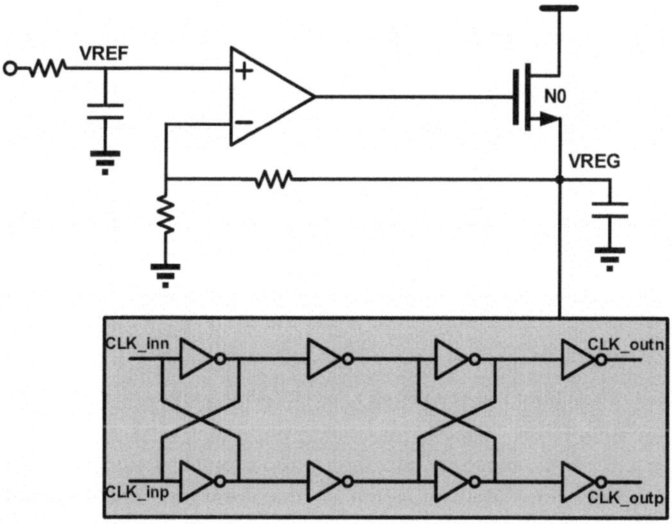

Fig. 6 Voltage regulator design

$$\text{Jitter}_{RMS} = \frac{2}{w_0^2} \int_{CDR_{BW}}^{+\infty} \Gamma(f) df \quad (1)$$

The main high-frequency noise of the clock path is the noise source from the power supply section. In this paper, a regulator design is added to the clock buffer in order to provide a stable power supply while avoiding the impact of clock path noise on other modules. As shown in Fig. 6, The regulator [10] consists of a high-gain, low-bandwidth error amplifier, followed by a thin-oxide NMOS source follower (N0). The loop can provide more than 40 dB of gain, which serves to stabilize the power supply. The amplifier is only used to set the correct DC level of VREG, so its bandwidth can be set very low to minimize power consumption.

3.3 PLL Circuit

In high-speed wireline transmitters, the power consumption of the PLL and the clock distribution network is a critical concern. In this work, a PLL is employed to generate a 2-channel 4-GHz output for serialization, with its block diagram presented in Fig. 7. To support arbitrary multi-rate operations while maintaining low power consumption and area-efficiency, a sigma-delta-modulation-based fractional-N PLL with a ring oscillator is adopted to achieve a wide-tuning range for multi-rate operations.

As the transmitter output jitter heavily relies on the quality of the transmitter clock jitter, a wide-bandwidth PLL is proposed to mitigate the VCO phase noise and

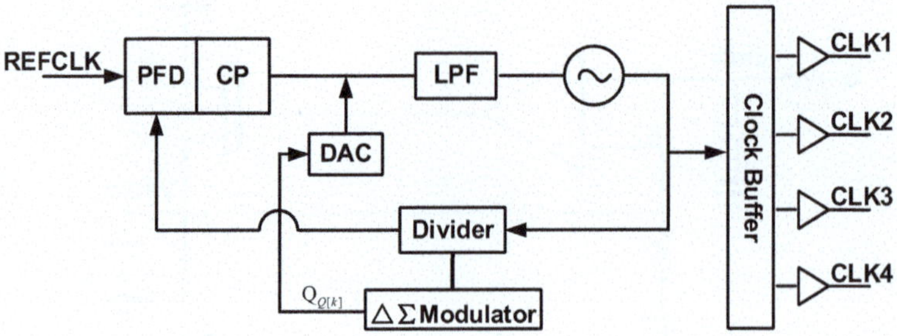

Fig. 7 PLL design

realize lower clock jitter. Nevertheless, the wide loop bandwidth has little impact on the suppression of the quantization noise from the $\sum - \Delta$[11]. To mitigate such noise, a capacitor DAC-based quantization noise module is proposed, as shown in Fig. 7. Once receiving the quantization residue from the Delta-Sigma, the DAC converts such noise into a current signal that exactly eliminates the quantization noise transmitted by the charge pump. This structure realizes an output range of 1 to 16 Gb/s, while simultaneously solving the quantization noise problem of the $\sum - \Delta$.

As shown in Fig. 7, the quantization noise $Q_Q[k]$ generated by the sigma-delta modulator will be injected into the PLL loop through the charge pump, resulting in the charge noise of $Q_Q[k]$, as shown in Eq. (2). The purpose of the spur cancellation circuit is to pass the residual quantization noise through the DAC to generate an inverse $-Q_Q[k]$, and remove the noise before it enters the LPF.

$$Q_Q[n] = I_{cp} T_{VCO} \sum_{k-n_0}^{n-1} e_Q[k] \tag{2}$$

The voltage-controlled oscillator (VCO) in this study employs a two-stage differential ring oscillator design. Although the ring oscillator falls short of the LC oscillator in terms of noise performance, it offers the advantages of small area and high portability, while the two-stage differential ring oscillator exhibits a relatively high resistance to common mode noise [12]. The structure of the two-stage differential ring oscillator is depicted in Fig. 8. Each differential tube of the oscillator is a pseudo-differential logic comprising a pair of main buffer stages and a pair of cross-coupled inverters. According to the Barkhausen criterion [13], the two-stage differential ring oscillator is required to provide a phase shift of 90 degrees at the oscillation frequency for each stage. The phase shift of each stage is contributed by both the main buffer stage and the cross-coupled inverter, with the overall phase shift increasing with the ratio between the cross-coupled inverter size and the main buffer stage size. For the overall phase shift to be greater than 90 degrees, this ratio should be at least 0.6. The variation in process, supply voltage, and temperature (PVT) affects both the main buffer stage and the cross-coupled inverter. However, this ratio remains relatively stable, thereby making the structure less susceptible to PVT variations [14, 15].

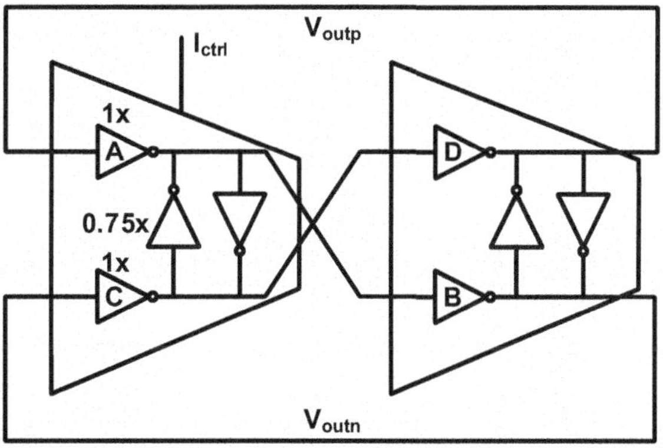

Fig. 8 VCO design

4 Measurement Results

Figure 9 shows the die micrograph of the transmitter. Two transmitter channels and one shared PLL are fabricated in 40 nm CMOS process, the overall chip area is 2 × 3 mm^2. To reduce the parasitic capacitance and inductance from the chip to the package, a package form of flip welding is employed in this paper (the package is customized according to the distribution of bumps on the layout). The silicon area of the analog front end (excluding the LVDS receiver, DSP, and PLL) is 316 × 152μm^2. The transmitter consumes 60.8 mW from 1.1/2.5-V supplies, giving an energy efficiency of 3.8 pJ/b. The contribution of the 40:1 MUX, the Driver, the MCG, and the clock distribution are 18.8 mW, 35.7 mW, and 6.3 mW respectively.

The clock test was configured in 1010 mode. It should be noted that the output clock's phase noise includes contributions from not only the PLL itself but also the buffers and output drivers. The test consisted of two parts: frequency domain and time domain. As shown in Fig. 10, the in-band noise was controlled to be below −90 dBc/Hz. However, some noise glitches appeared in the band, causing deterministic jitter in the time domain. This behavior was primarily due to the irrational factors of the charge pump. Nevertheless, this issue occurred within the tracking bandwidth of the CDR and did not significantly impact the final SerDes transmission bit error rate.

Figure 11 shows the eye diagram at the scope without any channel de-embedding. The figure shows a total jitter TJ of 56 ps at the output, with random jitter (RJ) of 1 ps RMS (14 ps peak-to-peak for 10^{-12} bit error rate) and deterministic jitter (DJ) of 42 ps. From the system point of view, under the BER of 10^{-12} at the receiving end, the eye-opening is greater than 0.5UI, which meets the design requirements in the SerDes standard (the specification requires greater than 0.3UI).

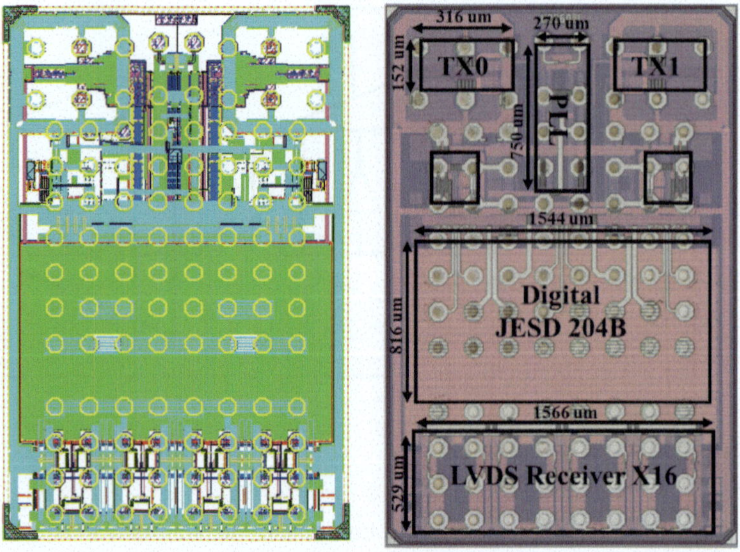

Fig. 9 Layout and die micro-photograph of the proposed transmitter

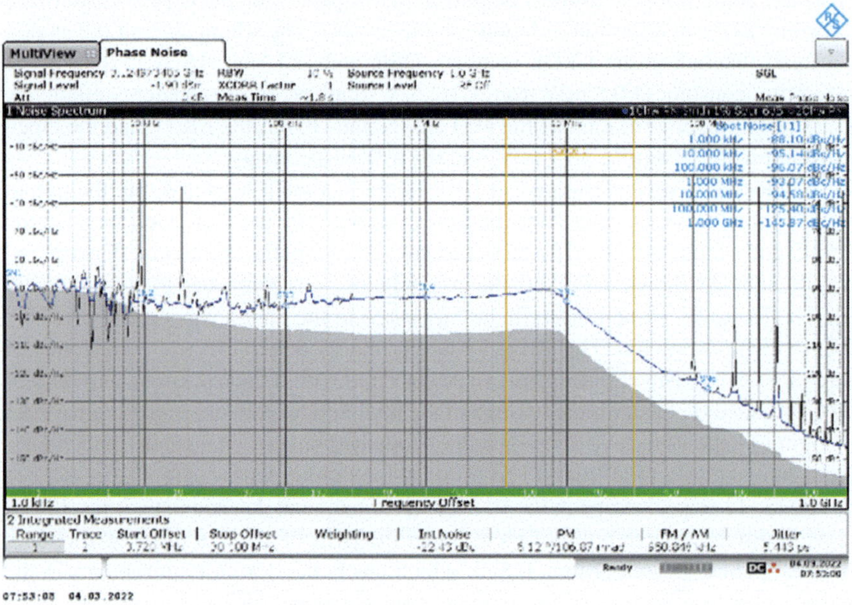

Fig. 10 Phase noise of the PLL output clock

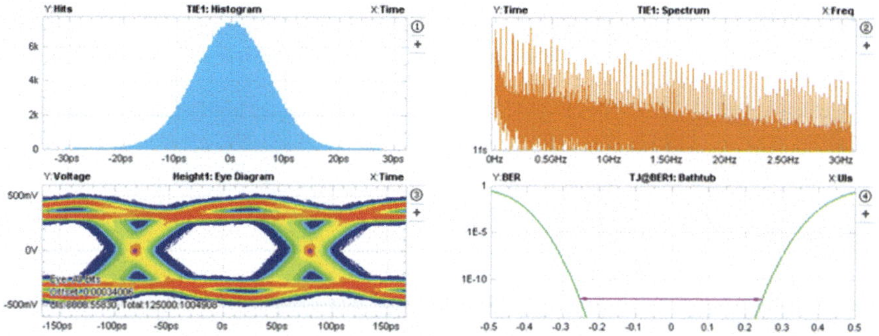

Fig. 11 Measured TX output eye

5 Conclusion

As a conclusion, this paper presented a low-power 16-Gbps SST transmitter with a voltage regulator structure to mitigate power supply noise interference. The SST structure effectively reduces the power consumption of the transmitter, while the sigma-delta fractional PLL enables good performance across a wide range of data rates. Fabricated in 40-nm CMOS, the transmitter consumes 60.8 mW from 1.1/2.5-V supplies, resulting in an energy efficiency of 3.8 pJ/b. The proposed transmitter has demonstrated excellent performance and is a promising solution for high-speed communication systems.

Acknowledgements This work was supported by the National Laboratory of Science and Technology on Analog Integrated Circuit (Grants No. JCKY2022210C006).

References

1. Wu Z, Zhang C, Li F, Z Wang (2016) High speed serial interface transceiver controller based on JESD204B. In: 2016 14th IEEE international new circuits and systems conference (NEWCAS) Vancouver BC, pp 1–4
2. Celik F, Akkaya A, Tajalli A, Burg A, Leblebici Y (2019) JESD204B compliant 12.5 Gb/s LVDS and SST transmitters in 28 nm FD-SOI CMOS. In: 2019 15th conference on Ph.D research in microelectronics and electronics (PRIME), pp 101–104
3. Ghafarian H et al (2021) A 9-bit 45 mw 0.05 mm2 source-series-terminated dac driver with echo canceller in 22-nm cmos for in-vehicle communication. IEEE Solid-State Circuits Lett 4:10–13
4. Kim J et al (2015) A 16-to-40 Gb/s quarter-rate NRZ/PAM4 dual-mode transmitter in 14 nm CMOS. In: Proceedings of IEEE international conference solid-state circuits (ISSCC), pp 1–3
5. "Serial Interface for Data Converters - JESD204C (Revision of JESD204B.01 January 2012)", December 2017 JEDEC solid state technology association 2017 Arlington VA, pp 22201–2107
6. Ding L, Jin J, Zhou J (2022) A 16/32Gbps dual-mode SerDes transmitter with linearity enhanced SST Driver. IEICE Trans Fundam Electron Commun Comput Sci E105(11):1443

7. Peng P-J et al (2018) A 50-Gb/s quarter-rate voltage-mode transmitter with three-tap FFE in 40-nm CMOS. In: Proceedings of IEEE 44th European solid-state devices and circuits conference (ESSCIRC), pp 174–177
8. Celik F et al (2021) A 32-Gb/s PAM4 SST transmitter with four-tap FFEUsing high-impedance driver in 28-nm FDSOI. IEEE Trans Very Large Scale Integr (VLSI) Syst 29(6):1132–1140
9. Bassi M et al (2016) A high-swing 45 Gb/s hybrid voltage and current mode PAM-4 transmitter in 28 nm CMOS FDSOI. IEEE J Solid-State Circuits 51(11):2702–2715
10. Kwon D-H (2020) A low-power 40-Gb/s pre-emphasis PAM-4 transmitter with toggling serializers. IEEE Trans Circuits Syst II Exp Briefs 67(3):430–434
11. Huang X, Zeng K, Rhee W, Wang Z (2019) A noise and spur reduction technique for $\Delta\Sigma$ fractional-N bang-bang PLLs with embedded phase domain filtering. In: 2019 IEEE international symposium on circuits and systems (ISCAS), pp1–4
12. Tang F et al (2020) A 32-step phase-compensated spread-spectrum RF-PLL with 19.44-dB EMI reduction and 10-fs extra RMS jitter. IEEE Trans Microw Theory Techn 68(4):1564–1575
13. Sachdeva P, Aggarwal A (2016) Design of CMOS ring VCO for PLL based frequency synthesizer. Int J Electr Electron Eng Telecommun 5(2):73–79
14. Anjana R, Kumar Somkuwar A (2012) Minimizing the sub threshold leakage for high performance CMOS circuits using novel stacked sleep transistor technique. Int J Electr Electron Eng Telecommun 1(1): 76–82
15. Kumar R (2015) Reduce power consumption for digital CMOS circuits using DVTs algoritham. Int J Electr Electron Eng & Telecommun 1(1): 376–383

The manufacturer's authorised representative in the EU is Springer Nature Customer Service Centre GmbH, Europaplatz 3, 69115 Heidelberg, Germany. If you have any concerns regarding our products, please contact ProductSafety@springernature.com

Printed and bound by CPI Group (UK) Ltd, Croydon, CR0 4YY
01/10/2026
02225926-0002